T0258047

Advanced Earthquake-Resistant Structures

Advanced Earthquake-Resistant Structures

Edited by **Bruno Crump**

New York

Published by NY Research Press,
23 West, 55th Street, Suite 816,
New York, NY 10019, USA
www.nyresearchpress.com

Advanced Earthquake-Resistant Structures
Edited by Bruno Crump

International Standard Book Number: 978-1-63238-014-2 (Hardback)

Contents

Preface

This book focuses on earthquake-resistant structures which can withstand major earthquakes. It comprises of research works contributed by various experts and researchers in the field of earthquake engineering. The book provides an overview of latest developments and advances related to earthquake-resistant structures. The book discusses seismic-resistance design of masonry and reinforcement of concrete structures and safety measurements, strengthening and rehabilitation of existing structures against earthquake loads. It also covers topics dedicated to seismic bearing capacity of shallow foundations, seismic behavior and retrofits of infilled frames, and also provides case studies related to seismic damage estimation in Mexico and vulnerability of buildings in western China. The book should be useful to graduate students, researchers and practicing structural engineers.

This book has been the outcome of endless efforts put in by authors and researchers on various issues and topics within the field. The book is a comprehensive collection of significant researches that are addressed in a variety of chapters. It will surely enhance the knowledge of the field among readers across the globe.

It is indeed an immense pleasure to thank our researchers and authors for their efforts to submit their piece of writing before the deadlines. Finally in the end, I would like to thank my family and colleagues who have been a great source of inspiration and support.

Editor

Compound Stochastic Seismic Vulnerability Analysis and Seismic Risk Evaluation for Super Large Cable-Stayed Bridge

Feng Qing-Hai[1,2], Yuan Wan-Cheng[1] and Chih-Chen Chang[3]
[1]State Key Laboratory of Disaster Reduction in Civil Engineering,
Tongji University, Shanghai
[2]CCCC Highway Consultants CO., Ltd. Beijing
[3]Department of Civil and Environmental Engineering, Hong Kong University of
Science and Technology, Clear Water Bay, Kowloon, Hong Kong,
China

1. Introduction

Previous seismic disasters indicate bridge structures are the most vulnerable component of the road transportation system in seism, such as the Haiti earthquake and Tohoku earthquake. Bitter lesson of bridge damage leads to the development of seismic analysis theories for bridge such as seismic vulnerability analysis.

Generally, seismic vulnerability is the probability of different damage in different seismic levels, which combines the intensity measure of seism with damage index for bridge structure. Methods of seismic vulnerability analysis include expert vulnerability analysis, experience vulnerability analysis and theory vulnerability analysis. Nevertheless, only seismic randomness has been taken into account in most seismic vulnerability analysis (e.g. Shinozuka et al. 2000; Kiremidjian et al. 1997; Basoz et al. 1997; Yamazaki et al. 2000). Apart from seismic randomness, bridge structure parameters are also stochastic, such as material properties, bridge geometry, boundary condition and so on, which cause the randomness of structure seismic response. Therefore, seismic vulnerability for bridge structure should be determined by both the randomness of seism and that of structure parameters. Due to the complexity of bridge structure system, it is very difficult to gain the analytic solution of seismic response for stochastic structure. Since the stochastic numerical simulation is time consuming and inefficient, this method is suspended at the threshold of thought.

Now, based on traditional seismic vulnerability analysis method, Artificial Neural Network (ANN), Monte Carlo (MC) technologies combining with Incremental Dynamic Analysis (IDA) and PUSHOVER method, Compound Stochastic Seismic Vulnerability Analysis (CSSVA) method is developed to take both randomness of material and that of seism into account from the point of total probability, which not only gives full play to the ANN, MC, IDA and PUSHOVER,but also increases the the efficiency of analysis greatly (Feng Qing-hai 2009).

Generally, the damage of bridge might lead to more serious results and secondary damage than that of road, and the seismic risk level of bridge determines that of the whole road transportation system. Due to the above mentioned reasons, more and more scholars pay attention to the bridge seismic risk evaluation.

Safety is relative, risk is absolute. So far, bridge structure risk evaluation is only limited to that of transportation, maintenance, management and so on. Since paper or reports for seismic risk evaluation for bridge structure performance is few, the only essential aspect for bridge seismic risk evaluation by those paper or reports is the seismic risk according to the damage probability of a determined bridge structure (e.g. Furuta et al. 2006; Hays et al. 1998; Padgett et al. 2007). However, that is unilateral. The truth is the randomness of material and that of seism exist at the same time. Moreover, the bridge damage probability according to the seismic vulnerability is gained within seism happening. In order to overcome this shortage, a method of seismic risk evaluation based on IDA and MC is presented, which has taken the difference of site type, randomness of time, space and intensity into consideration. It has indeed reflects the seismic risk situation within any years.

So, by the aforementioned method for seismic vulnerability and risk assessment, more real performance state and seismic risk lever are gained, which are good for design, maintenance and earthquake insurance of long span bridge.

2. Method of compound stochastic seismic vulnerability analysis

2.1 Basic theory

The Damage Index(DI) of bridge is affected by many uncertainties such as randomness of parameters of structure, seism and so on, which makes it a complicated process in gaining the seismic vulnerability curve. The capability of bridge structure could be expressed in the form of $R(M,G,C,......)$, M, G, C stand for variety of material, size and calculation methods, respectively. At the same time, the seismic response could be expressed as $P(IM)$. So, DI is the systemic combination of $R(M,G,C,......)$ and $P(IM)$, which is shown as following.

$$DI = f\left(R(M,G,C,......),P(IM)\right) \qquad (1)$$

Clearly, DI is also variable.

Easy to see, the calculation of DI for bridge structure is divided into several parts, the first is the statistic for capability of bridge structure itself, the second is the statistic for seismic response of bridge structure, and the final is the combination of them. The flow chart is shown in Figure 1.

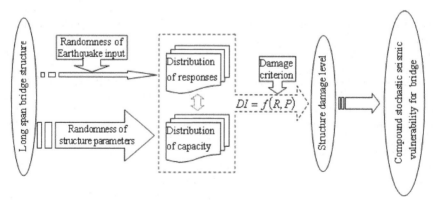

Fig. 1. Flow Chart for Compound Stochastic Seismic Vulnerability Analysis

2.2 Statistic for capability of bridge structure

Statistic for capability of bridge structure is analyzed based on method of PUSHOVER combining ANN and MC. The analysis process is shown in the following. Radial Basic Function Neural Network (RBFNN) is adopted in this method according to the conclusion of reference (Feng Qing-hai 2007).

1. Main parameters which affect the capability of bridge structure most are analyzed. Distributions for each parameter are determined, too.
2. By the method of orthogonal design, A+B groups of finite element models of bridge structures are built.
3. The response of capability is derived by the method of PUSHOVER.
4. RBFNN is built, and trained by responses of A groups and checked by that of B groups.
5. Go on if the result of Step 4 is successful, or rebuild RBFNN from Step 4.
6. Plenty of responses of capability are simulated by inputting large number of structure model parameters generated by the method of MC and RBFNN.
7. The characteristic of capability of bridge structure is gained by statistics on all those responses of capability.

The analysis flow chart is shown in Figure 2.

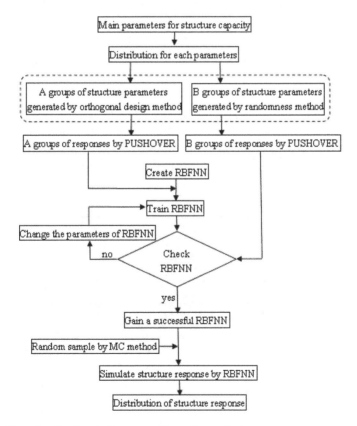

Fig. 2. Flow Chart for Stochastic Structure Capacity analysis

2.3 Statistic for seismic response of bridge structure

Seism is a ground motion of great randomness which might happen at anytime and anywhere. Therefore, it is very necessary to analyze the seismic response from the view of probability. While, the variability of bridge structure itself has a small effect on the seismic response when the randomness of both structure and ground motion are considered (Hu Bo 2000). In order to simplify the calculation method and reduce calculation time, only seismic randomness is taken into account in the statistic for seismic response of bridge structure in this paper.

Statistic analysis of seismic response is performed based on the method of IDA under multi-earthquake waves. The analysis process is shown in the following.

1.   The finite element model of bridge structure is built.
2.   Multi-earthquake waves are selected, and scaled to different intensities by scale factors (SF).
3.   IDA is performed. Plenty of seismic responses are gained.
4.   The characteristic of seismic responses of bridge structure is gained by statistic analysis on all those seismic responses.

The analysis flow chart is shown in Figure 3.

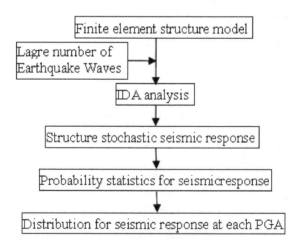

Fig. 3. Flow Chart for Stochastic Seismic Response Analysis

2.4 Methodology of CSSVA for bridge structure

Based on the theories mentioned in above sections, the distribution characteristic of capability and seismic response of bridge structure, CSSVA is performed combining with ANN-MC technology. The analysis process is shown in the following.

1. Get the distributions of capability and seismic responses of bridge structure, respectively.
2. Generate adequate numbers of capability and seismic responses by the method of MC and orthogonal design based on the distribution characteristic got in Step 1.
3. Two RBFNN are built, one for capability and the other for seismic response, then trained and checked by the data gained in Step 2.
4. Go on if the result of check is successful, or rebuild RBFNN from Step 3.
5. A large number of DI are gained according to Equation 1.
6. Select the damage criterion.
7. Compare DI with the damage criterion. The probability of damage is calculated at each IM.
8. The seismic vulnerability curves of bridge structure are drawn.

The analysis flow chart is shown in Figure 4.

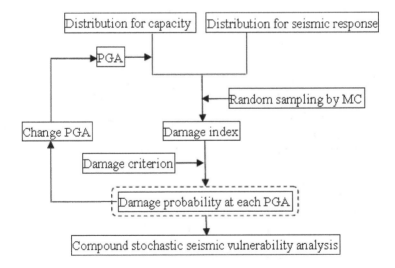

Fig. 4. Flow Chart for Compound Stochastic Seismic Vulnerability Analysis

2.5 Effect on seismic vulnerability by compound stochastic

Based on CSSVA, some examples are performed. For the limited space of paper, only part of results is drawn to display the effect on seismic vulnerability by compound stochastic. Seismic vulnerability curve of determination structure and that of compound stochastic are shown in Figure 5.

Easy to see, when structure stochastic are neglected, seismic vulnerability curves, which are gained by the method of two order spline curve fittings, are folded. Meanwhile, the curve deviate to the vulnerability point distinctly. By comparison, vulnerability curves of compound stochastic are smooth, and are getting through every vulnerability point. For the two curves, the PGA for beginning damage is the same. However, when damage probability is less than 80%, vulnerability value of determination structure is bigger than that of compound stochastic. When damage probability ranges from 80% to 95%, the contrary is the case.

In all, vulnerability curve of determination structure spreads around that of compound stochastic. Besides, the structure stochastic indeed affects the seismic vulnerability, although not too enormous. Compound stochastic reflects the seismic vulnerability of bridge structure better.

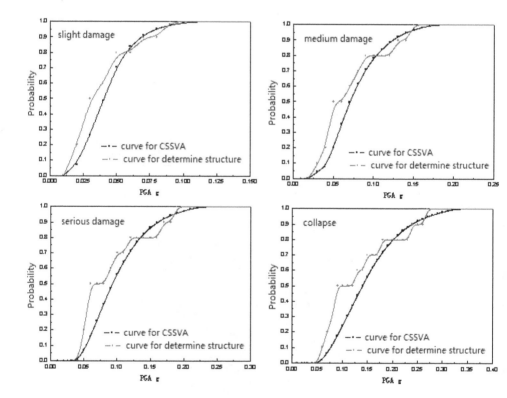

Fig. 5. Effect on Seismic Vulnerability by Compound Stochastic

3. Seismic risk probability evaluation for long span bridge

3.1 Concept for probability evaluation of seismic risk

Bridge seismic risk analysis involves two aspects, namely the probability of event occurrence and that of event consequence. The consequence mainly describes the damage severity and has no relationship with structure analysis. Risk analysis, usually said, is the probability of event occurrence without considering the consequence.

Seismic risk probability evaluation for bridge is one of the basis of risk analysis, which also involves two aspects, namely seismic risk and structure seismic vulnerability. Seismic risk probability evaluation for bridge could be expressed as following: the probability of different damage states with consideration of earthquake dangerous. It indicates that the reliability of bridge is threatened by both seism and structure seismic vulnerability during the same determined period.

Seism is regarded as a risk event, the sign is H. According to structure reliability thought, limit state function could be expressed as following (Alfredo H-S. ANG & Wilson H.Tang, 2007).

$$Z = R - S \tag{2}$$

in which, Z is the performance function, R is the comprehensive resistance, S is comprehensive response on structure induced by risk events. Then, the damage probability is expressed by expression as following.

$$P = P(R < S) = \int_R^\infty f(S)dS \tag{3}$$

in which, $f(S)$ is the probability density function of comprehensive effect induced by risk events. Clearly, S is closely related to H. The $f(S)$ should be expressed in the form of $f(S,H)$, namely the joint density function.

$$f(S,H) = f(S|H)f(H) \tag{4}$$

in which, $f(S|H)$ is the conditional probability density function for different damage states under given risk event H. $f(H)$ is the probability density function of risk event H. So, $f(S)$ is expressed as following.

$$f(S) = \int_{-\infty}^{+\infty} f(S|H)f(H)dH \tag{5}$$

then, the expression combining Equation 5 and Equation 3 is as following.

$$P = p(R < S) = \int_R^{+\infty}[\int_{-\infty}^{+\infty} f(S|H)f(H)dH]dS = \int_0^{+\infty}[\int_R^{+\infty} f(S|H)dS]f(H)dH \tag{6}$$

$$P = \int_0^{+\infty} F_S(H)f(H)dH \tag{7}$$

briefly,

$$P = \int_0^{+\infty} F_S(H)f(H)dH \tag{8}$$

in which, $F_S(H)$ stands for $\int_R^{+\infty} f(S\,|\,H)dS$.

3.2 Method of seismic risk probability evaluation for bridge based on IDA-MC
In order to avoid complicated calculation, IDA is applied in association with MC. The specific procedure is interpreted at the following sections.

3.2.1 Probability distribution of seismic intensity and according PGA
Seismic dangerous analysis is the basis of seismic risk probability evaluation, which reflects the probable maximum effect of seismic damage for a bridge in a district within determined coming period. The effect could be depicted in many ways. Since bridges designed in China are based on the design criterion of seismic intensity, probability distribution of seismic intensity introduced in reference (Gao Xiao-wang et al. 1986) is adopted as the seismic dangerous, namely:

$$F_{III}(x) = \exp(-(\frac{\omega - x}{\omega - \varepsilon})^K)$$
(9)

in which, ω is the upper limited value of intensity, usually equals to 12. ε usually satisfied with the equation of $1 - e^{-1} = 0.632$. K is form parameter. From the point of engineering application, K equals to the value of intensity according to the seismic probability of 10%.

If the probability distribution of seismic intensity for 50 years is determined, then the probability distribution in any limited period could be expressed as the following.

$$F_i(i) = [F_T(I)]^{t/T} = [\exp(-(\frac{\omega - i}{\omega - \varepsilon})^K)]^{t/T} = \exp(-\frac{t}{T}(\frac{\omega - i}{\omega - \varepsilon})^K)$$
(10)

According to the existing seismic records, PGA according to a determined intensity are of big discreteness. In order to be convenient for calculation, A has the relationship to I as following.

$$A = 10^{(I \cdot Log\,2 - 0.01)}$$
(11)

in which, A and I is the value of PGA and seismic intensity, respectively. The unit for A is gal. Here, it is very necessary to illuminate that, A is a continuous value only for the necessary of statistic.

3.2.2 Method and steps
1. To build a finite element analysis model of bridge structure, and choose enough seismic waves.
2. IDA is performed. Seismic response according to every wave is recorded to form IDA curve.
3. Determine design reference period (prior to a small one) and the probability distribution.
4. According to MC, a great lot of I are generated.
5. According to Equation 11, the same number of A are gained based on Step 4.

6. Based on A, corresponding seismic responses are gained from the IDA curve.
7. Structure damage probability is gained by statistics based on damage criterion.
The analysis procedures are shown in Figure 6.

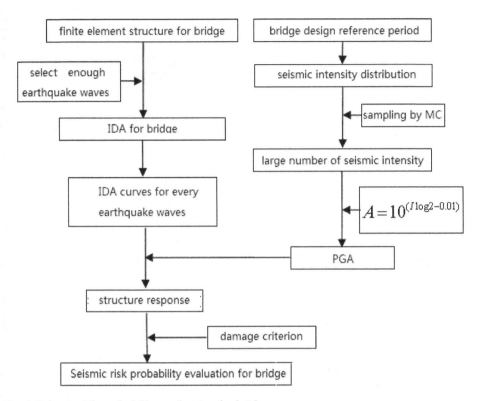

Fig. 6. Seismic risk probability evaluation for bridge

3.3 Example analysis
3.3.1 Analysis model for floating system cable-stayed bridge
According to reference (Yan Hai-quan and Wang Jun-jie 2007), floating cable-stayed bridge could be simplified to a main tower with a lumped mass on the top of main tower, and the tolerance of results are acceptable. In this section, simplified models are adopted and only longitudinal cases are analyzed. The refined finite element model for main tower is built by OpenSees as shown in Figure 7, main tower is divided into 3 parts, namely upper tower, middle tower and low tower. In consideration of the strengthening at the upper tower by steel pile casting, the upper tower is regarded elastic. Fiber element is adopted in middle tower and low tower.

Main tower is 300m in height, 90m for upper tower, 150m for middle tower and 60m for low tower, respectively. Tower is divided into 30 elements of 10 meters in vertical direction, and every element and joint is numbered from 1 to 30 and 1 to 31, respectively. The bridge site belongs to 3rd type, the seismic intensity is 8 and design period is 50 years.

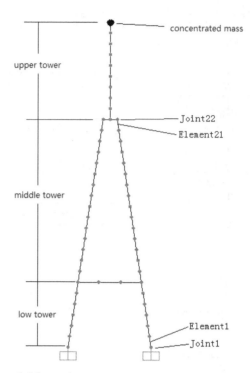

Fig. 7. Finite element model for main tower

3.3.2 Damage criterion

Double control damage criterion put forward by Park and Ang are adopted, which is shown as following:

$$DI = f\left(\delta_m, \delta_u, Q_y, \int dE_h, \beta\right) = \frac{\delta_m}{\delta_u} + \frac{\beta}{\delta_u Q_y}\int dE_h \tag{12}$$

in which, the signal and suggestion value are illuminated particularly in reference (Fan Li-chu and Zhuo Wei-dong 2001). The relationship between damage levels, damage state and DI described by Park, Ang and Wen (H. Hwang et al. 2001) is shown in Table 1.

Damage level	Damage character	Park-Ang DI
1 no damage	Some slight fracture at local part	$DI < 0.1$
2 slight damage	Slight fracture distributed widely	$0.1 \le DI < 0.25$
3 medium damage	Serious fracture or spall partly	$0.25 \le DI < 0.4$
4 serious damage	Concrete crushed or steel break	$0.4 \le DI < 0.8$
5 collapse	collapse	$DI \ge 0.8$

Table 1. Damage Character and Damage Index at Each Level

3.3.3 IDA for main tower

Only longitudinal case is studied. IDA is performed with PGA being scaled from 0.1g to 1.0g with step of 0.1g. Seismic responses such as moment, curvature, hysteretic energy are recorded. Representative response distributions along main tower are described in Figure 8, 9,10, respectively..

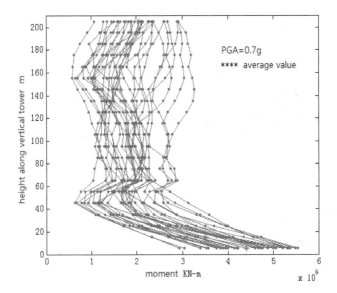

Fig. 8. Moment distribution along vertical tower

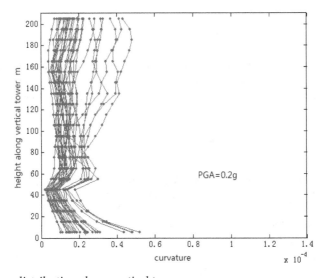

Fig. 9. Curvature distribution along vertical tower

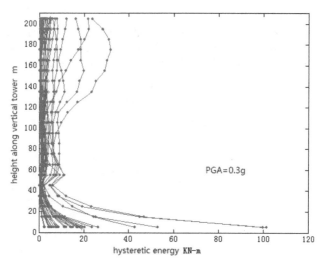

Fig. 10. Hysteretic energy distribution along vertical towe

3.3.4 Stochastic sampling for PGA

According to reference (Ye Ai-jun 2002) combining the site type, design period and seismic intensity, K is determined as following:

$$1 - 0.1 = \exp(-(\frac{12-8}{12-6.45})^K)$$ (13)

According to Equation 13, the K=6.87. Then, the seismic intensity probability distribution within t years is

$$F_t(i) = \exp(-\frac{t}{50}(\frac{12-i}{5.55})^{6.87})$$ (14)

When t=1, and according to the method of MC, 50000 random intensity value are generated, numbered $Rand(1)$ to $Rand(50000)$. Then, the according PGA numbered from $RandPGA(1)$ to $RandPGA(50000)$ which reflect the seismic random from time, space and intensity are gained as following.

$$RandPGA(i) = 10^{(Rand(i)\log(2)-0.01)}$$ (15)

3.3.5 Bridge structure seismic risk probability evaluation within 1 year

Based on the stochastic PGA and IDA curve, responses for every element in any stochastic PGA are gained, the total number is 560000. Based on the damage criterion, the damage time for every element in different damage levels are accounted. For example, the damage number for slight damage at the bottom of main tower is 7649. Then the damage probability is

$$\frac{7649}{560000} \times 100\% = 1.37\%$$ (16)

Damage probability distribution along main tower in different damage level is shown in Figure 11(a). Easy to see, in different damage levels, the damage probability is different, especially damage probability of the bottom is larger than that of the others, there are two main reasons for this phenomenon as following:

Firstly, the main tower is the most important component of large long span bridge, any damage could cause devastating damage for the bridge. So in the design, main tower is conservatively designed. Not only stiffness but also strength is strengthened.

Secondly, when the bottom element begins to be damaged, the upper elements are protected.

Based on the thought of structural reliability, the main tower is a series of structure, so the total damage probability is the biggest one in different damage levels.

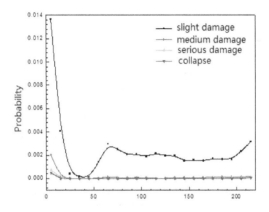

(a) Damage probability distribution along vertical tower at different states

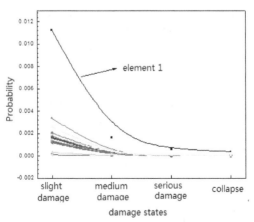

(b) Element damage probability at different damage states

Fig. 11. Element damage probability at different damage states

3.3.6 Bridge structure seismic risk probability evaluation within any years

With different years, the seismic risk is different, and so does the seismic intensity distribution. According to the method as in Section 3.3.5, seismic risk probability from 10 to 100 years with interval of 10 years are calculated, as shown in Figure 11(a).

In order to get a reasonable cognition, risk levels are shown in Table 2. Combining figure 11(a) with Table 2(Basoz, Nesrin, Kiremidjian & Anne S, 1997), the log curve for main tower seismic risk is shown in Figure 11(b).

Grade	Risk Description	Probability Range	Intermediate Value
1	Very unlikely	<0.0003	0.0001
2	Impossible	$0.0003-0.003$	0.001
3	Occasional	$0.003-0.03$	0.01
4	Possible	$0.03-0.3$	0.1
5	Very likely	>0.3	1

Table 2. Risk Probability Description

Easy to see from Figure 11(b), for main tower, within 1 year, slight damage and medium damage is impossible, and serious damage will never happen. Within other years, especially over 30 years, slight damage is possible, medium damage and serious damage will be occasional, and collapse will never happen.

4. Conclusion

Compound stochastic seismic vulnerability analysis for bridge structure is presented, in which double damage criterion is adopted. The effect of compound stochastic on seismic vulnerability are analyzed, which proves that, compound stochastic indeed has little effect on seismic vulnerability. The results show that compound stochastic seismic vulnerability curve reflects the actual situation.

Based on method of MC and results of seismic vulnerability analysis for large long span cable-stayed bridge, seismic risk analysis evaluation for bridges is performed, which has taken difference of site type, randomness of time, space and intensity into consideration. Seismic risk situation within any years for bridge structure is measured by using the experience of situation description for bridge risk evaluation.

5. Acknowledgements

This research is supported by Ministry of Science and Technology of China under Grant No. SLDRCE 09-B-08, Kwang-Hua Fund for College of Civil Engineering, Tongji University and the National Science Foundation of China under Grant No.50978194 and No.90915011, and Science and technology project of Ministry of Transport, China.

6. References

Alfredo H-S. Ang & Wilson H.Tang.(2007). Probability concepts in engineering..

Shinozuka M., Feng, M.Q., Lee, J. & Naganuma, T. (2000). Statistical Analysis of Fragility Curves[J]. *Journal of Engineering Mechanics, ASCE*, Vol. 126, No.12, (December, 2000), pp. 1224–1231

Kiremidjian A. S. & Bosöz, N. (1997). Evaluation of Bridge Damage Data from Recent Earthquakes. *NCEER Bulletin*, Vol. 11, No. 2, pp. 1-7

Basoz, Nesrin, Kiremidjian & Anne S. (1997). Risk Assessment of Bridges and Highway Systems from the Northridge Earthquake. *Proceedings of the National Seismic Conference on Bridges and Highways: "Progress in Research and Practice"*, pp. 65-79, Sacramento, California, USA

Yamazaki, F., Motomura, H. & Hamada, T. (2000). Damage Assessment of Expressway Networks in Japan based on Seismic Monitoring. *Proceeding of the 12th World Conference on Earthquake Engineering*, pp. 0551, Upper Hutt, New Zealand

Feng Qing-hai (2009). Study on Seismic Vulnerability and Risk Probability Analysis of Super-Large Bridge[D]. Tongji University, Shanghai, China

Furuta H., Katayama H. & Dogaki M. (2006) · Effects of Seismic Risk on Life-cycle Cost Analysis for Bridge Maintenance. *Proceedings of the 4th International Conference on Current and Future Trends in Bridge Design, Construction and Maintenance*, pp. 22-33

Hays W.W · (1998). Reduction of Earthquake Risk in the United States : Bridging the Gap between Research and Practice · *IEEE Transactions on Engineering Management*, Vol.45, No.2, (May 1998), pp. 176-180

Padgett, Jamie E, DesRoches & Reginald (2007). Bridge Functionality Relationships for Improved Seismic Risk Assessment of Transportation Networks. *Earthquake Spectra*, Vol.23, No.1, (February 2007), pp. 115-130

Feng Qing-hai, Yuan Wan-cheng (2007). Comparative Study on BP Neural Network and RBF Neural Network in Performance Evaluation of Seismic Resistance for Pier Columns[J]. *Structrual Engineers*, Vol.23, No.5, pp. 41-47

Hu Bo (2000) · Study on the Probabilistic Seismic Design Method for Bridges[D] · Tongji University, Shanghai, China

Gao Xiao-wang, Bao Ai-bin (1986). Determination of Anti-Seismic Level by Probabilistic Method[J]. *Journal of Building Structures*. Vol.3, No.2, pp. 55-63.

Yan Hai-quan; Wang Jun-jie (2007). A Tower Model for Seismic Response Prediction of Floating Cable-stayed Bridge in Longitudinal Direction[J]. *Journal of Earthquake Engineering and Engineering Vibration*, , Vol.27, No.4, (February 2007), pp. 80-86.

Fan Li-chu, Zhuo Wei-dong (2001). Seismic Design for Ductility of Bridge[M]. *Beijing: China Communications Press*, 2001

California Office of Emergency Services, Vision 2000: Performance Based Seismic Engineering of Buildings, *Structural Engineers Association of California*, Sacramento, CA, 1995

H. Hwang, J.B.Liu & Y.H.Chiu (2001). Seismic Fragility Analysis of Highway Bridges, *Mid-Ameirica Earthquake Center Technical Report*, MAEC-RR-4 Project, 2001

Ye Ai-jun (2002). Bridge Seismic[M]. *Beijing: China Communication Press*, 2002

2

Recent Advances in Seismic Response Analysis of Cylindrical Liquid Storage Tanks

Akira Maekawa
Institute of Nuclear Safety System, Inc.
Japan

1. Introduction

Japan is in a seismically active area and experiences many damaging earthquakes with loss of life. In 1995, the Hyogoken-Nanbu earthquake caused major destruction in Kobe City and in 2011 the Great Eastern Japan earthquake and tsunami caused major destruction in the Pacific coast areas of northeastern Japan. Additionally, recent relatively less destructive earthquakes include the 2003 Tokachi-Oki earthquake, the 2004 Niigataken Chuetsu earthquake, the 2005 Miyagiken-Oki earthquake, the 2007 Noto-Hanto earthquake, and the 2007 Niigataken Chuetsu-Oki earthquake.

In particular, the Great Eastern Japan, Noto-Hanto, Miyagiken-Oki, and Niigataken-Chuetsu-Oki earthquakes occurred near nuclear power facilities and have been accompanied by enhanced public concern for seismic safety of nuclear plants. Within the Japanese national government, the Nuclear Safety Commission revised the *Regulatory Guide for Reviewing Seismic Design of Nuclear Power Reactor Facilities* (Nuclear Safety Commission of Japan, 2006) in 2006. This revised *Regulatory Guide* required seismic safety design of buildings, structures and equipment for larger seismic motions. In addition, seismic probabilistic safety assessment (seismic PSA) (American Nuclear Society, 2007; Atomic Energy Society of Japan, 2007) was urged, for which accurate evaluation techniques for seismic response of equipment installed in the nuclear power plants were needed.

Large cylindrical liquid storage tanks in nuclear power plants are classified as equipment requiring high seismic safety because many are containers storing cooling water used in normal plant operation and in accidents. Their seismic evaluation is done on the basis of the *Technical Codes for Aseismic Design of Nuclear Power Plants* (Japan Electric Association [JEA], 2008) published by the Japan Electric Association. The seismic evaluation methods used in the conventional seismic design of the tanks (Kanagawa Prefecture, 2002; High Pressure Gas Safety Institute of Japan [KHK], 2003; Architectural Institute of Japan [AIJ], 2010) such as the *Technical Codes* examine the bending vibration mode (beam-type vibration) which mainly affects the seismic resistance of the tanks, but they do not consider high order vibration modes (oval-type vibration) which are excited in the tank wall by large vibrations and cause oscillation patterns that look like petals of a flower. Therefore, it is necessary to reveal the influence of oval-type vibration on vibration characteristics and seismic safety and to consider the vibration in the seismic design of the tanks (Japan Society of Civil Engineers, 1989). However, research on oval-type vibration has only been of academic interest, including reports on fluid-structure interaction which causes oval-type vibration (Japan Society of Mechanical Engineers, 2003) and nonlinear behavior of oval-type vibration (Chiba, 1993). Though analysis techniques such as finite element methods are available as seismic evaluation methods at present, numerical seismic analysis of the tanks considering oval-type vibration has not been established because advanced techniques such as fluid-structure interaction analysis and nonlinear dynamic structure analysis are needed to simulate the oval-type vibration behavior.

In addition, capacity to resist buckling is an important evaluation item in seismic deign of cylindrical liquid storage tanks. Buckling is a dangerous mode for tanks which drastically lowers their structural strength (proof force) and collapses their geometries. In the ultimate buckled state and post-buckling, the cylindrical liquid storage tanks are deformed largely and display nonlinear inelastic behavior. Therefore, it is desirable to take into account the nonlinear inelastic dynamic behavior when evaluating seismic safety of the tanks. The conventional seismic design of tanks (Kanagawa Prefecture, 2002; KHK, 2003; JEA, 2008; AIJ, 2010) uses evaluation equations for static buckling derived from static buckling tests and the assumption of a linear response. However, the evaluation equations have not been validated sufficiently from the viewpoint of the dynamic liquid pressure effect in tanks subjected to seismic motions. Though a few dynamic experiments and development of numerical methods for buckling of cylindrical liquid storage tanks were done in the past, the developed numerical methods could simulate the experimental results only qualitatively (Fujita et al., 1992; Toyoda et al., 1997).

As described above, the conventional seismic design assumes linear behavior of the tanks and does not include nonlinear behavior in post-buckling. However, it is necessary to develop accurate seismic response analysis methods for the cylindrical liquid storage tanks to ensure seismic safety and conduct accurate seismic PSA for mega earthquakes. Therefore, an accurate dynamic analysis method to evaluate dynamic nonlinear behavior of the cylindrical liquid storage tanks subjected to seismic motions was proposed and validated by the dynamic experiment in this chapter. The research was done for the liquid storage tanks installed in nuclear power plants such as refueling water tanks and condensate water tanks. In this chapter, previous studies are overviewed and then sequential research findings on the dynamic analysis method are summarized.

First, the seismic damage modes of the cylindrical liquid storage tanks are explained briefly. Especially buckling modes caused by earthquakes are introduced. Secondly, the vibration behavior of the tanks is explained. Thirdly, previous studies are overviewed with regard to vibration characteristics and seismic evaluations. Special focus is given to the seismic response analysis and dynamic buckling evaluation. Fourthly, research studies concerned with oval-type vibration are summarized. Finally, the author's study regarding dynamic nonlinear analysis method for seismic response of the cylindrical liquid storage tanks is described and the method is shown to be suitable for actual tanks based on comparison with experimental results.

2. Seismic damage modes of cylindrical liquid storage tanks

Many typical examples of seismic damage modes of cylindrical storage tanks have been reported and many seismic damage analyses (see for example, (Fujii et al., 1969; Rahnama &

Morroe, 2000; Suzuki, 2008)) have been conducted. The damage modes of the tanks from the above analysis results are summarized as follows:
1. Buckling in the side walls
2. Failure of the tank roofs and their junctions
3. Sliding and lifting
4. Local fracture on the bases of the tanks and uneven settlement
5. Failure of anchor bolts
6. Cracking of annular parts of the base plate

The buckling modes of the side walls of tanks include shear buckling and bending buckling. The bending buckling includes diamond buckling and elephant foot bulges. These buckling modes are associated with geometry parameters of the tanks such as height to radius ratio and radius to thickness ratio. Figure 1 shows the relationship between the buckling modes and geometry parameters. Shear buckling occurs for small ratios of height to radius and bending buckling predominantly occurs for large ratios. Shear buckling is caused by shear force and brings about many large diagonal wrinkles in the center of a tank side wall. A typical example of diamond buckling is shown in Fig. 2. Diamond buckling is one of the bending buckling modes caused by the bending moment and it is generated on the base of a tank. When the buckling occurs, the cross section at the buckling region bends inward and has many wrinkles. Because the deformation is drastic, the structural strength (proof force) of the tanks decreases suddenly. Diamond buckling became widely know after it occurred in many wine storage tanks in the 1980 Greenville-Mt. Diablo earthquake. The elephant foot bulge is another bending buckling mode. This buckling mode was widely seen in the 1964 Alaska mega earthquake, the 1971 San Fernando earthquake and the 1994 Northridge earthquake and can cause spill incidents of liquid in the tanks through crack penetration. A typical example of the elephant foot bulge is shown in Fig. 3 (Ito et al., 2003). In the elephant foot bulge, the buckling cross section expands outward in a ring and the structural strength (proof force) decreases relatively gently through a gradual increase of the expansion. The occurrence condition of diamond buckling and the elephant foot bulge depends on the circumferential stress due to the internal pressure in the tanks, that is, hoop stress (Akiyama et al., 1989). The former occurs when the hoop stress is smaller and the latter occurs when the hoop stress is larger. In the 1995 Hyogoken-Nanbu earthquake, many observations of diamond buckling and elephant foot bulges were made in cylindrical liquid storage tanks.

The failure of tank roofs and their junctions is mainly caused by sloshing. This occurred in the 1964 earthquakes in Niigata and Alaska. More recently, the roofs and junctions of some petroleum tanks failed in the Kocaeli earthquake in Turkey and in the Chi-Chi earthquake in Taiwan, both in 1999. In Japan, a few petroleum tanks also failed in the 2003 Tokachi-Oki earthquake. In all three of these earthquakes, the floating roofs were damaged and fires broke out.

The sliding of tanks and lifting of base plates were observed in the 1964 Alaska earthquake. Local fracture on the bases of tanks, uneven settlement and failure of anchor bolts occurred in the 1995 Hyogoken-Nanbu earthquake. In the 1978 Miyagiken-Oki earthquake, cracking of annular parts of base plates occurred in petroleum tanks and stored petroleum leaked out.

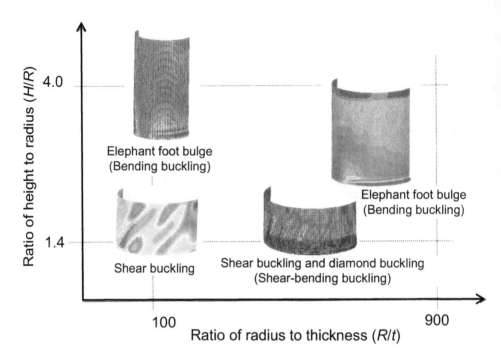

Fig. 1. Buckling modes of cylindrical tanks and geometry parameters.

Fig. 2. Typical mode of diamond buckling.

Fig. 3. Typical mode of elephant foot bulge. (Ito et al., 2003)

3. Classification of vibration behaviour in cylindrical liquid storage tanks

The vibration modes of the cylindrical liquid storage tanks are classified roughly into sloshing and bulging. The sloshing represents vibration of the free liquid surface and the bulging represents vibration of the tank structure. In large cylindrical tanks, shell vibration occurs because the side wall is relatively thinner compared to the radius length, which is regarded as a cylindrical shell. The modes of bulging including shell vibration are generally distinguished using the axial half wave number m and circumferential wave number n. The modes with $m \geq 1$ and $n = 1$ are the beam-type vibration and the modes with $m \geq 1$ and $n \geq 2$ are the oval-type vibration. Typical examples of the vibration modes are shown in Fig. 4 (Fujita & Saito, 2003). The figure shows the vibration modes of cylindrical tanks which are free on the top and rigid on the bottom; it is easy for readers to understand these modes of oval-type vibration. However, actual tanks have the vibration condition which is rigid on the top because of their fixed roofs.

In general linear analysis, vibration modes with $n \geq 2$ are not excited when a perfectly axisymmetric cylindrical shell such as a tank is vibrating. However, the oval-type vibration actually occurred in the vibration experiment using reduced models of cylindrical tanks (Kana, 1979; Fujita et al., 1984; Maekawa et al., 2010). Additionally it is not possible to say that the influence of oval-type vibration on seismic load of the tanks can be ignored (Clough et al., 1979).

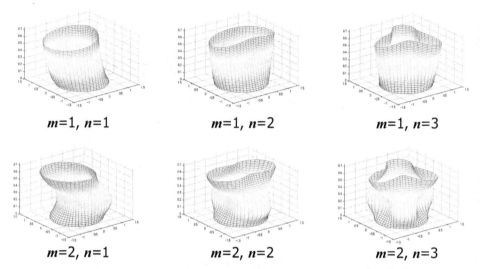

$m=1, n=1$ $m=1, n=2$ $m=1, n=3$

$m=2, n=1$ $m=2, n=2$ $m=2, n=3$

Fig. 4. Typical vibration modes of tanks: m, axial half wave number; n, circumferential wave number. (Fujita & Saito, 2003)

4. Overview of previous studies on vibration characteristics and seismic resistance of cylindrical liquid storage tanks

In cylindrical liquid storage tanks, the liquid and the structure compose the coupled vibration system between fluid and structure and show complex vibration characteristics. Jacobsen (1949) and Werner and Sandquist (1949) were the first to study the influence of the contained liquid on dynamic behavior of cylindrical tanks. Their studies focused on only hydrodynamic behavior of the contained liquid assuming rigid containers. In the 1950s, the National Aeronautics and Space Administration (NASA) actively investigated the vibration behavior of rocket fuel tanks and then developed many empirical and analytical methods of sloshing and coupled vibration between fluid and structure (Abramson, 1966). The studies on sloshing developed as a specific field and their overview was done by Ibrahim et al. (2001). As for seismic evaluation of cylindrical liquid storage tanks, Housner (1957) proposed seismic response analysis method in 1957 which became known as Housner's theory and it has been adopted in some seismic design guidelines for cylindrical tanks. In this theory, the dynamic liquid pressure is calculated based on two separate pressures due to the horizontal inertia force of liquid and sloshing of the free liquid surface when cylindrical liquid storage tanks are subjected to seismic motions. However, tanks are assumed to be rigid bodies. Since Housner's proposal, many researchers have studied seismic evaluation methods. Veletsos and Yang (1976) and Fischer and Rammerstorfer (1982) proposed seismic response analysis methods of cylindrical liquid storage tanks assuming flexible structures for them by simplifying the vibration mode shape of the tanks. Moreover, Fujita (1981) also proposed seismic response analysis method in which tank structure was modeled by finite elements, the liquid pressure in tanks was expressed using velocity potential theory and series solution and coupled vibration between tank structure and contained liquid was considered. Ma et al. (1982) proposed an analysis method in which both tank structure and contained liquid were modeled by finite elements. The studies mentioned above focused on development of vibration response analysis methods of tanks subjected to seismic motions. Rammerstorfer et al. (1990) and Shimizu (1990) summarized research prior to 1990 on seismic evaluation methods of cylindrical liquid storage tanks. In their reviews, a few topics including bending vibration (beam-type vibration) and sloshing were covered, but not oval-type vibration. At present the behavior of oval-type vibration including the influence on seismic resistance of tanks remains unclarified.

Since the 1990s the issues of seismic study of cylindrical liquid storage tanks have shifted to dynamic buckling behavior. Though the buckling problem of cylindrical tanks was studied in the mid 1960s (NASA, 1968; Timoshenko et al., 1964), only the structural strength of an empty cylinder was investigated. The buckling problem of cylindrical liquid storage tanks subjected to seismic motions was in the limelight in the 1964 Alaska earthquake and in the 1971 San Fernando earthquake. In the 1980s Yamaki (1984) studied buckling behavior of cylindrical tanks filled with liquid systematically. In those studies, static buckling behavior was examined using small cylindrical tanks. Various dynamic buckling experiments were performed using different sizes of cylindrical tanks (Shin & Babcock, 1980; Niwa & Clough, 1982), and at the same time the occurrence mechanism of the buckling in the tanks was researched theoretically. In Japan, some dynamic buckling experiments were conducted using large scale models of cylindrical liquid storage tanks (Katayama et al., 1991; Tazuke et al., 2002). These experiments focused on proving tests of actual tanks, investigating the occurrence mechanism of the elephant foot bulge, and reflecting the experimental results into the seismic design. The experimental objects were vessels of fast breeder reactor and tanks for liquid natural gas. Toyoda and Masuko (1996), Fujita et al. (1990) and Akiyama (1997) studied the dynamic buckling of vessels and tanks vigorously and systematically, and Akiyama's proposal was adopted in a few guidelines for industrial tanks and vessels. However, the dynamic buckling behavior was not always clarified and Fujita et al. (1990) only examined the influence of oval-type vibration on seismic behavior of tanks. The numerical analysis methods, which could take into account fluid-structure interaction due to the contained liquid, were developed to simulate these experimental results but were not sufficient because of limited analysis conditions. Ito et al. (2003) conducted a dynamic buckling experiment on tall cylindrical liquid storage tanks such as the refueling water tanks installed in nuclear power plants. In their experiment, the ultimate state of the cylindrical liquid storage tanks was clarified but no numerical simulation development was discussed. Nowadays, seismic PSA should be performed as part of the seismic evaluation in nuclear power facilities. This requires that accurate failure modes and structure strength must be grasped for the ultimate state of equipment such as tanks. Therefore, it is urgent that the dynamic buckling simulation methods be established.

On the other hand, oval-type vibration has been studied from the standpoint of vibration engineering and the vibration behavior of a cylindrical shell, that is shell theory, has been investigated. Shell theory was studied from the 1920s and was systematically summarized by researchers such as Timoshenko and Woinowsky-Krieger (1959). The study on oval-type vibration of cylindrical liquid storage tanks was started in the 1960s as investigations of vibration and dynamic instability of a cylindrical shell partially filled with water. Chiba et al. (1984, 1986) conducted vibration tests using small cylindrical containers and undertook a theoretical study using Donnell's shell theory. More recently, Amabili (2003) investigated nonlinear response of oval-type vibration using Donnell's nonlinear thin shell theory vigorously. The contents of studies in the research field were reviewed by Amabili and Païdoussis (2003). However, these studies focused on nonlinear vibration characteristics and the instability region of oval-type vibration. Recently Maekawa et al. (2006) and Maekawa and Fujita (2007) clarified the influence of oval-type vibration on the resonance frequency and amplitude ratio of beam-type vibration in cylindrical tanks by vibration tests using a test tank. Their findings indicate that it is necessary to simulate oval-type vibration accurately in the seismic response analysis of cylindrical liquid storage tanks.

5. Present status of dynamic analysis of cylindrical liquid storage tanks in Japan

Cylindrical liquid storage tanks have a simple cylindrical geometry, but generate complex vibration behavior. In large tanks, the vibration behavior is a coupled system between the liquid and the tank structure because the tank walls are relatively thin and deformable. In the coupled system, various vibration modes simultaneously occur during earthquakes because the natural frequencies of most of the modes are in the exciting frequency range of the earthquakes. In the seismic design of tanks, the beam-type vibration which is the bending vibration mode is evaluated as the vibration mode which dominantly affects seismic response of the tanks. However, the vibration behavior of tanks is complex because oval-type vibration also actually occurs, which is a high order vibration mode and oscillation occurs with a petal-like pattern in the side wall. Highly advanced simulation techniques are needed to analyze such a complex vibration behavior. In conventional seismic evaluation methods, structural analysis is conducted assuming a hydrodynamic pressure distribution of liquid in the tanks as the load distribution with added mass based on velocity potential theory. However, the accuracy of time history analysis using conventional analysis methods is not good. Actually, the time history analysis of the hydrodynamic pressure distribution for successive fluid-structure interaction analysis is needed to simulate the behavior of beam-type vibration and oval-type vibration accurately.

The conventional seismic evaluation of cylindrical liquid storage tanks (Kanagawa Prefecture, 2002; KHK, 2003; JEA, 2008; AIJ, 2010) is performed assuming the tank structure responses linearly and using evaluation equations for static buckling obtained on the basis of static buckling tests. However, an accurate buckling analysis method with dynamic effects and consideration of fluid-structure interaction has never been proposed. Though attempts were made to simulate dynamic buckling behaviors of cylindrical tanks (Fujita et al., 1992; Toyoda et al., 1997), accurate simulation methods were not realized.

Most recently Maekawa and Fujita (2009) proposed an accurate seismic response analysis and a dynamic buckling analysis for cylindrical liquid storage tanks. The proposed dynamic analysis method using explicit finite element analysis was validated by a dynamic seismic experiment which indicated that the analysis accuracy was good for evaluation of seismic response (vibration response and sloshing) and buckling behavior of cylindrical liquid storage tanks.

6. Proposed dynamic analysis method

The conventional seismic response analysis for cylindrical liquid storage tanks (Kanagawa Prefecture, 2002; KHK, 2003; JEA, 2008; AIJ, 2010) is appropriate as a conservative seismic design method but it is not appropriate for grasping actual phenomena during earthquakes.

Maekawa and Fujita (2008) have proposed a nonlinear seismic response analysis method for cylindrical liquid storage tanks. In this method, the tank structure is three-dimensionally modeled using shell elements which can consider geometric nonlinearity (Belytschko et al., 1984), the contained liquid is modeled using solid elements which can calculate fluid behavior according to Euler's equation, and the fluid-structure interaction between the contained liquid and the tank structure is calculated by the arbitrary Lagrangian-Eulerian method (ALE method) (Hirt et al., 1974). Maekawa and Fujita (2008) compared analysis results obtained by their proposed method with experimental results and demonstrated that the nonlinear vibration behavior caused by the influence of oval-type vibration as well as the behavior of oval-type vibration could be simulated by the proposed method.

Maekawa and Fujita (2009) also applied their proposed method to dynamic buckling analysis. A nonlinear inelastic analysis and a large deformation analysis were conducted considering material nonlinearity to simulate plastic deformation and response in the post-buckling. The characteristics of the proposed method and the conventional dynamic buckling analysis method (Fujita et al., 1992; Toyoda et al., 1997) are compared in Table 1. As summarized there, the seismic response by the proposed method can be simulated more accurately and realistically.

Figure 5 shows an analysis model of a cylindrical tank used to validate the proposed method. This is a detailed finite element model of a test tank used for the dynamic buckling experiment presented in later sections. The numbers of nodes and elements are 424,460 and 410,482, respectively. Figure 6 shows a photo of the test tank.

In the analysis model, the Lagrange elements are used for the structure part and the Euler elements are used for the fluid part in the model. The Belytschko-Lin-Tsay shell elements (Belytschko et al., 1984) based on Mindlin theory are used for modeling the tank structure. The shell elements can consider geometric nonlinearity and shear deformation. The shear deformation is formulated by rotating a cross section vertical to the center of a shell element surface and the nonlinear constitutive equations are expressed by only linear terms using an approach to remove the rigid body motion. For the present analysis case, a 200-kg weight on the tank top was assumed as an added mass rigidly fixed on the top of the model. The imperfection distribution from a perfect circle of the cylinder obtained by profile measurement was set in the model. The material constants used in the analysis are listed in Table 2. The stress-strain relationship of aluminum alloy used in the cylinder was obtained by a material test and was approximated by multiple lines for the elasto-plastic input condition. The material constants of steel and polycarbonate were nominal values.

The fluid part is modeled by solid elements which use Euler's equation as the fundamental equation. Using Euler elements, the fluid moves from one element to another element to reproduce the pressure change due to vibration. For the present analysis case, the properties of water were set for the 95% water level of the model and the remaining 5% part was assumed as empty because air and void elements were used. The ALE method is used for the fluid-structure interaction analysis.

In the analysis, static pressure of the liquid in the tank and the dead weight of the tank structure should be considered. Hence, the load balance in the static state was calculated by loading gravity force to the model before the dynamic analysis. The mass-proportional damping with 100% damping ratio was assumed to obtain the stationary solution quickly. Next, the analysis model was impacted using a triangular wave, the free vibration of the model was excited, and then the natural frequency of the primary beam-type vibration was obtained by frequency analysis of the response. Based on its natural frequency of 30.45 Hz, 31 Hz was chosen as the excitation frequency. Finally, the dynamic buckling simulation was performed using sinusoidal waves with 31 Hz as input in which the liquid pressure change was calculated successively. The input acceleration 2.5 G was chosen based on the experimental conditions described in the next section. The analysis conditions are summarized in Table 3. Time history analysis was conducted by an explicit integral method using the finite element analysis code LS-DYNA (Livermore Software Technology Corporation, 2003).

From the viewpoint of convergence, it is difficult to solve all problems when large deformation problems such as buckling are solved by an implicit method which often is used in the general finite element method. Therefore, the explicit method was applied to the seismic response analysis, which is a long time analysis, though the method has been used for only extremely short time analyses such as impact analysis until now. This was possible due to the rapid progress which has been made in computer calculation performance.

Items	Proposed method	Conventional method (Fujita et al., 1992; Toyoda et al., 1997)
Behavior of water contained in the tank	Contained water modeled by solid elements; fluid-structure interaction calculated sequentially	Contained water not modeled; added mass considered.
Oval-type vibration	All modes considered	Higher order modes not considered
Sloshing	Analyzable	Non-analyzable
3D modeling	3D model for both axisymmetric and non-axisymmetric structures using shell elements	Axisymmetric model using axisymmetric elements
Solving method	Explicit method	Newmark β method (Implicit method)
Geometry nonlinearity	Rigid body motion removed to allow use of a simple constitutive equation with linear terms for efficient calculation and better analysis accuracy	Complicated constitutive equation with high order nonlinear terms used for better analysis accuracy
Material nonlinearity	Elasto-plastic model using multiple line approximation of stress-strain relationship	Elasto-plastic model using multiple line approximation of stress-strain relationship

Table 1. Comparison of the proposed and conventional dynamic buckling analysis methods

Materials \ Items	Young's modulus (MPa)	Poisson's ratio	Density (kg/m³)	Bulk modulus (MPa)
Aluminum alloy (A5052) (cylinder)	69,420	0.33	2,680	–
Steel (flange and platform)	203,000	0.3	7,800	–
Polycarbonate (top plate)	1,960	0.3	11,900	–
Water (in the tank)	–	–	1,000	2,200

Table 2. Material constants for analysis

Items	Analysis conditions
Structure	Model using nonlinear shell elements
Water	Model using solid element with Euler's equation
Air	Model using solid elements with voids
Weight	Model using mass elements
Fluid-structure interaction analysis	ALE method
Time-history analysis	Explicit time integration method
Damping	Rayleigh damping (3% at resonance point)
Excitation wave	Sinusoidal waves similar to those in the experiment
Excitation acceleration	2.5 G
Excitation frequency	31 Hz
Excitation direction	Horizontal (0°–180°)

Table 3. Analysis conditions

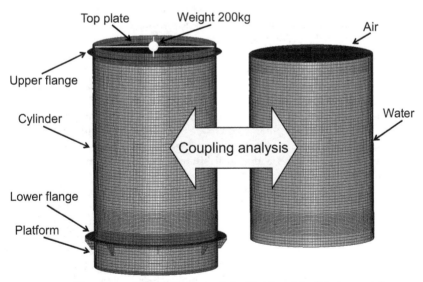

Structure part (shell elements)   Fluid part (solid elements)

Fig. 5. Analysis model of 1/10 reduced-scale cylindrical liquid storage tank.

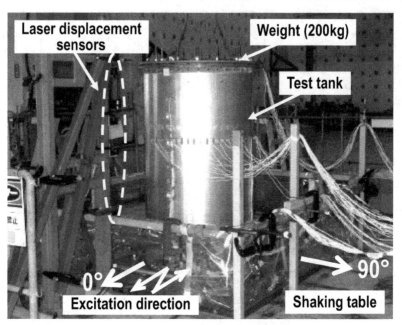

Fig. 6. Test tank of 1/10 reduced-scale cylindrical liquid storage tank.

7. Dynamic buckling experiment

The dynamic buckling experiment (Maekawa et al., 2007) of the test tank shown in Fig. 6 was conducted using a large shaking table setup to validate the proposed method. The test tank was a 1/10 reduced-scale model of the large cylindrical storage tanks installed in nuclear power plants such as condensate water tanks which are classified as equipment requiring the highest seismic safety. A 200-kg doughnut-shaped weight was put on the top of the tank to amplify the response of the tank during shaking.

The dimensions and measurement locations of the test tank are shown in Fig. 7. The test tank was an aluminum alloy cylinder with a wall thickness of 1 mm. The cylinder was fixed using steel flanges at both ends. A polycarbonate transparent board was used for the roof of the tank to look in the tank. The cylinder was made by shaping an aluminum alloy plate and using TIG welding. The shape imperfection from a perfect circle was measured by optical digital profilometry and the imperfection distribution was set in the analysis model. For the important geometry parameters in buckling, the height to radius ratio was 2.67 and radius to thickness ratio was 450. The test tank was fixed on the shaking table using a steel platform with a 2-mm wall thickness. The shear force and bending moment of the whole tank were estimated using strain generated in the platform during excitation. The behavior of beam-type vibration was measured using accelerometers on the top of the tank. Laser displacement sensors were put in the 0° direction at four heights of 290, 680, 870 and 1200 mm. The displacement changes of beam-type vibration and oval-type vibration were measured at the 1200-mm height and the other heights, respectively. In Fig. 7, many strain gauges were attached in the range from -18° to 138° at the 700-mm height. The mode shape of oval-type vibration was examined from the magnitude of strain in each position.

Occurrence of the oval-type vibration was observed by a video camera at the 90° position. The test tank was partially filled with water to the 95% level (to a height of 1140 mm) and it was excited horizontally between 0° and 180° using sinusoidal waves. The acceleration amplitude of the sinusoidal waves gradually increased and decreased at the start and end, respectively because other frequency components, except the excitation frequency, were not included. The natural frequency of primary beam-type vibration of the tank was examined by a frequency sweep test. The natural frequency of 27 Hz was chosen as the excitation frequency. Three exciting tests were conducted while changing the magnitude of the input acceleration.

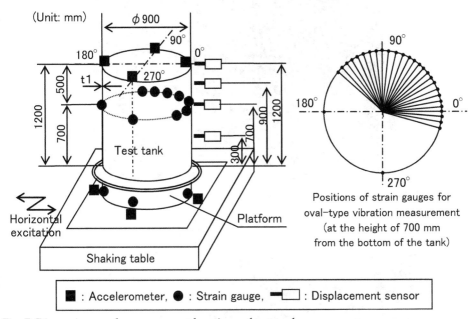

Fig. 7. Dimensions and measurement locations of test tank.

8. Dynamic buckling experiment results

Some distinct pictures expressing the dynamic behavior of the tank side wall are shown in Fig. 8. The vibration behavior of the wall is represented using lighted and shaded reflections from a spot light. The reflection of light in the center region of the cylinder was wide and narrow in the axis direction and indicated geometric change of the wall with time. Namely, oval-type vibration occurred and oscillation was large.

As shown in Fig. 9, plastic deformation was retained in the test tank after the experiment. The deformation in the side wall of the tank was probability caused by shear buckling and the deformed denting inward in the base was caused by bending buckling, specifically diamond buckling. Comparison of the maximum values of the shear stress and bending stress and the design allowance values indicated bending buckling occurred predominately because only the measured bending stress exceeded the design allowance value as shown in Table 4.

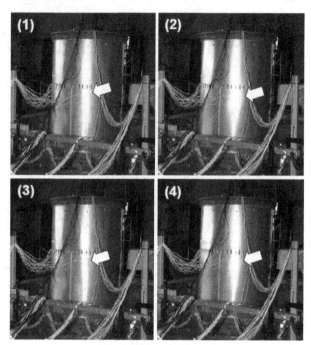

Fig. 8. Oval-type vibration occurring in the side wall of the test tank.

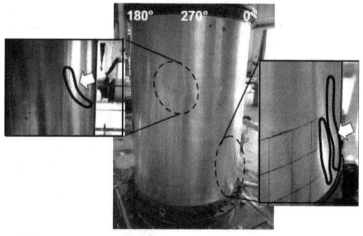

Fig. 9. Plastic deformation remaining in the tank wall after buckling.

	Experimental values (N/mm²)	Design allowance values (N/mm²)
Bending stress	97.18	47.02
Shear stress	33.45	54.65

Table 4. Comparison of buckling stresses

9. Results of proposed dynamic analysis

The time history analysis was conducted by the proposed method based on the analysis conditions in Table 3. Figure 10 shows distortion shape of the model after the analysis. The deformation factor was ten. Oval-type vibration occurred with large deformation of the wall. The bending buckling (diamond buckling) occurred with deformed denting inward in the base. These behaviors were similar to the experimental observation. Figure 11 shows the measured deformation shapes of a cross section of the test tank at a height of 700 mm. These shapes represent modes of oval-type vibration excited in the experiment before and after buckling. These modes were different, indicating the influence of buckling. Figure 12 shows the deformation shapes of a cross section of the model obtained by the analysis. It was found that the analytical shapes were similar to those in Fig. 11. Namely, the proposed dynamic analysis method was demonstrated to simulate the experimental results accurately. Figures 13 and 14 show the pressure distributions of liquid in the tank model obtained by the analysis at the corresponding times to Fig. 12. Figure 13 shows liquid pressure distributions of the cross section at a height of 700 mm and Fig. 14 shows the vertical section between 0° and 180°. The shape of the liquid pressure distributions in Fig. 13 corresponded to the mode shapes of oval-type vibration, for example, the regions are indicated by arrows. This indicates that the proposed method can simulate phenomena caused by the fluid-structure interaction in the tanks accurately. Consequently from the above results, the proposed dynamic analysis method is judged to be able to simulate oval-type vibration, liquid pressure behavior, buckling modes, and buckling response accurately.

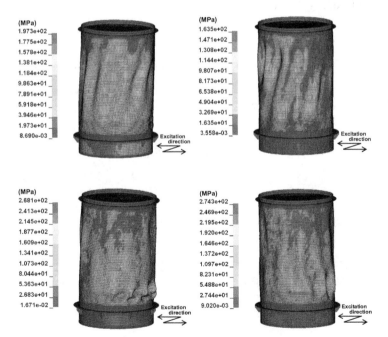

Fig. 10. Distortion shape and von Mises stress contour plots (displacement factor, 10) before (upper) and after (lower) buckling.

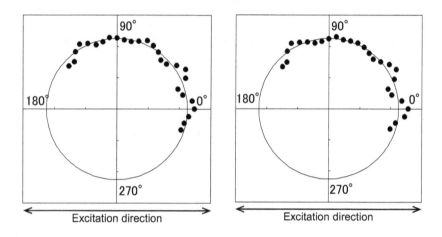

Fig. 11. Measured mode shapes of oval-type vibration in test tank at 700-mm height before (left) and after (right) buckling.

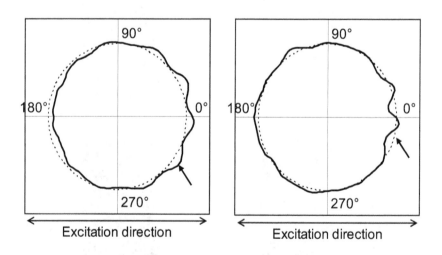

Fig. 12. Calculated mode shapes of oval-type vibration in tank model at 700-mm height before (left) and after (right) buckling.

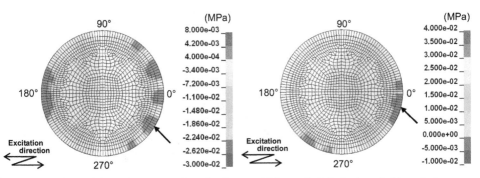

Fig. 13. Calculated liquid pressure distributions at 700-mm height before (left) and after (right) buckling.

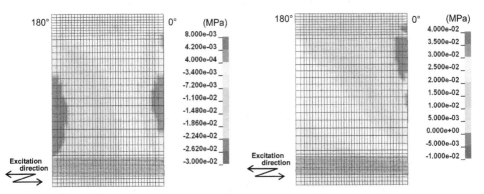

Fig. 14. Calculated liquid pressure distributions vertical section between 0° and 180° before (left) and after (right) buckling.

The mode shape of oval-type vibration excited was then discussed in detail using the proposed method. Figures 15 and 16 show the oval-type vibration behavior before and after buckling, respectively. The mode shape of oval-type vibration before buckling was specific and the circumferential wave number $n=12$ was dominant. On the other hand, the mode after buckling was unstable and non-uniform. This might be associated with the change of vibration characteristics of the tank model due to the development of buckling on the base of the tank model. As described above, the vibration behavior of cylindrical liquid storage tanks including oval-type vibration can be examined by numerical simulation using the proposed dynamic analysis method.

The dynamic behavior of the liquid pressure distribution also was discussed using the proposed method (see Figs. 13 and 14). In the liquid pressure distribution before buckling, a large dynamic liquid pressure occurred due to the large beam-type vibration. The change of liquid pressure due to oval-type vibration occurred locally. In the pressure distribution after buckling, the change of the dynamic liquid pressure due to beam-type vibration and oval-type vibration was not clear. The change of liquid pressure before and after buckling might also be attributed to the change of the vibration characteristics due to buckling.

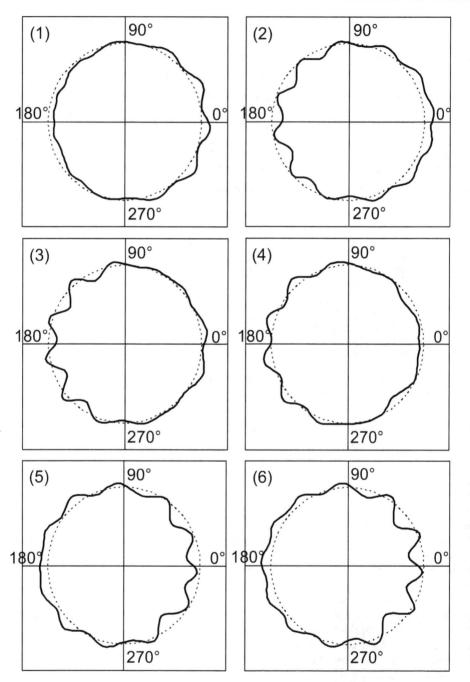

Fig. 15. Mode shape variation of oval-type vibration before buckling
(time course: from (1) to (6)).

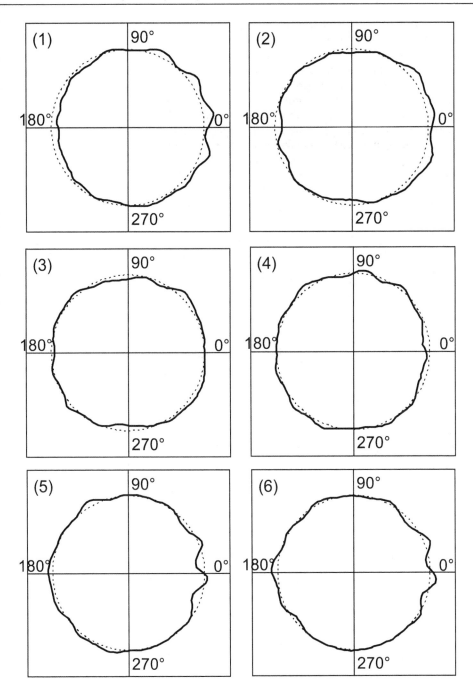

Fig. 16. Mode shape variation of oval-type vibration after buckling
(time course: from (1) to (6)).

The detailed analysis results of the dynamic liquid pressure distribution are shown in Figs. 17 and 18. The two figures show the vertical distribution of dynamic liquid pressure at the 0° position before and after buckling, respectively. Here, a qualitative discussion was done because the lines of the graphs were ragged which could be attributed to the roughness of the mesh division and inherent properties of the explicit method. The two figures suggested that the dynamic liquid pressure distribution was caused by beam-type vibration and the distribution shape was typical of that in thin-walled and flexible cylindrical liquid storage tanks. The distribution shapes before and after buckling differed. The difference was thought to be associated with the change of vibration characteristics of the tank due to buckling.

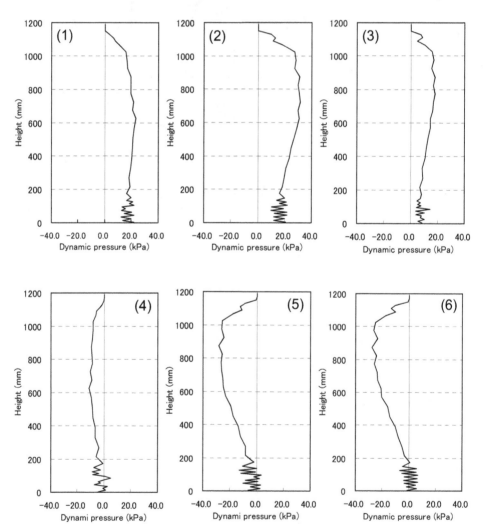

Fig. 17. Calculated vertical distributions change of dynamic liquid pressure at 0° position before bucking (time course: from (1) to (6)).

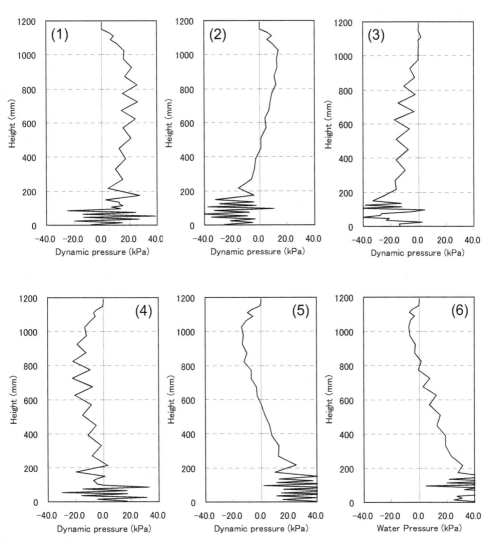

Fig. 18. Calculated vertical distributions change of dynamic liquid pressure at 0° position after buckling (time course: from (1) to (6)).

The acceleration and buckling loads obtained from the dynamic experiment and the analysis are summarized in Table 5. The shear force and bending moment agreed well between them. The load that worked on the tank was well estimated by the analysis. Hence, the proposed dynamic analysis method can estimate dynamic behavior and buckling behavior of liquid storage tanks accurately even though the seismic response of the tanks was unknown. The input acceleration in the analysis agreed with that in the experiment. This indicates that the input magnitude to increase buckling, that is, the magnitude of seismic motion, is predicable by the proposed dynamic analysis method.

In conclusion, it is demonstrated that the proposed dynamic analysis method can simulate the vibration response of cylindrical liquid storage tanks, the pressure behavior of the contained liquid and the buckling behavior accurately. Moreover, the buckling load and input magnitude to increase buckling can also be predicted accurately.

	Dynamic buckling experiment	Dynamic buckling analysis
Input acceleration (G)	2.37	2.61
Shear force (kN)	47.3	59.4
Bending moment (MN mm)	61.8	67.7

Table 5. Input excitation and associated base shear and base bending moment

10. Applicability to actual tank

The test tank and model described above are approximately a 1/10 reduced-scale model of an actual tank with uniform wall thickness. However, actual tanks are of many dimensions and shapes, and so research on larger tanks of different shapes would help assess the applicability of the proposed dynamic analysis method to actual tanks. Some dynamic buckling experiments using large-scale test tanks have been conducted. Ito et al. (2003) recently did a dynamic buckling experiment using a test tank similar to the refueling water storage tanks installed in nuclear power plants. The applicability of the proposed dynamic analysis method was assessed by using their experiment results as a benchmark (Maekawa & Fujita, 2010).

A photograph of their tank is shown in Fig. 19 (Ito et al., 2003). It was a 1/5 reduced-scale model. The tank, which had an internal diameter of 2200 mm, cylinder height of 2650 mm and water level of 2526 mm, had a non-uniform wall thickness with thickness decreasing in the height direction and was made of aluminum alloy. The height to radius ratio was 2.4 and the radius to average thickness ratio was 894. The experiment results showed the occurrence of oval-type vibration in the wall and bending buckling (elephant foot bulge) in the base; the type of bending buckling was different from that of the 1/10 reduced-scale test tank. Figure 20 shows the analysis model, which has 125,767 nodes and 131,712 elements. The Belytschko-Lin-Tsay shell elements (Belytschko et al., 1984) were used for the tank structure and solid elements with Euler's equation were used for the contained fluid. The coupling analysis between fluid and structure used the ALE method (Hirt et al., 1974) to estimate dynamic fluid-structure interaction. The free liquid surface was made by modeling the contained liquid and gas as solid elements in order to simulate sloshing. The same seismic wave as that used in the experiment was used as input. The magnitude of input acceleration was 2.7 Se (1600 Gal instantaneous maximum acceleration). In the experiment, buckling was caused and developed by the seismic wave of 2.7 Se.

The natural frequency of primary beam-type vibration of the analysis model was 16.6 Hz and that agreed with the experiment value of 16.8 Hz, showing the appropriateness of the model. The analytical result for the seismic wave is shown in Fig. 21. The side wall of the model was severely deformed inward, representing the occurrence of oval-type vibration. The wall in the base of the model locally came outward, showing the occurrence of elephant foot bulge. These results confirm that the proposed method can simulate the occurrence of oval-type vibration and the elephant foot bulge. In addition, Fig. 22 shows the simulation of sloshing of the contained liquid. The boundary between the liquid and air was heaved up, representing sloshing of the free liquid surface. Finally, these results demonstrate that the proposed method can also accurately simulate the vibration characteristics and buckling behavior of tanks with various shapes.

Fig. 19. Large test tank of 1/5 reduced-scale cylindrical liquid storage tank. (Ito et al., 2003)

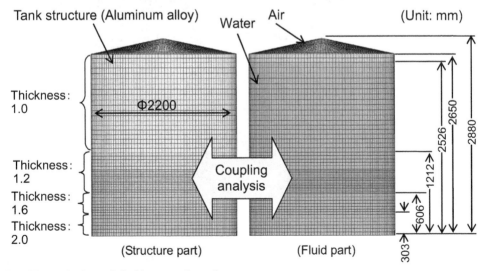

Fig. 20. Analysis model of large-scale tank.

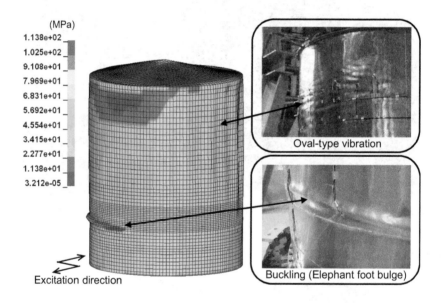

Fig. 21. Simulation of oval-type vibration and elephant foot bulge.

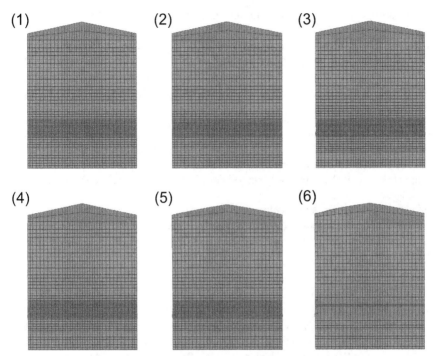

Fig. 22. Simulation of sloshing before (1) and during ((2) to (6)) excitation. The contained liquid is shown in red and the air in blue.

Comparison of analysis and experiment results for the 1/10 and 1/5 reduced-scale models show that the proposed method can simulate seismic response such as vibration response and sloshing and buckling behavior such as buckling mode and buckling load accurately. Consequently, the proposed dynamic analysis method is versatile and can be used to analyze many types of tanks. Additionally it is concluded that the method can adequately evaluate the seismic resistance of tanks such as their seismic safety margin (Maekawa et al., 2011).

11. Conclusions

The past and present research studies on seismic response analysis for cylindrical liquid storage tanks were reviewed. In addition, a new dynamic analysis method of seismic response for cylindrical liquid storage tanks which was developed by Maekawa and Fujita was introduced. The dynamic analysis method to evaluate seismic response (vibration and sloshing) and buckling was proposed and validated by the experimental results. The conclusions obtained are summarized as follows.

1. A new dynamic analysis method was proposed for evaluating seismic response and buckling behavior of cylindrical liquid storage tanks. In the method, the tank structure was three-dimensionally modeled by nonlinear shell elements which allowed geometric nonlinearity to be considered, the contained liquid was modeled by solid elements that calculated fluid behavior according to Euler's equation, and the fluid-structure

interaction between contained liquid and tank structure was evaluated by the arbitrary Lagrangian-Eulerian method.

2. In the proposed method, the explicit method was applied to the seismic response analysis, which is a long time analysis, though the method had been used for only extremely short time analysis such as impact analysis until now. The time history analysis results using the explicit method could reproduce the vibration and buckling behavior accurately.

3. A dynamic buckling experiment was conducted by using the 1/10 reduced-scale model of an actual tank. The experiment values were compared with the analysis results by the proposed method. The analysis and experiment results were in good agreement, especially with vibration response, buckling mode and buckling load. The results demonstrated that the proposed method could accurately simulate the seismic response of the tank.

4. The experiment and analysis values were compared for a 1/5 reduced-scale model. The proposed method was judged to be versatile and applicable to many types of tanks.

5. It was concluded that the proposed method could adequately evaluate the seismic resistance of tanks such as their seismic safety margin.

12. Acknowledgment

Some figures and photos were provided by Prof. Fujita of Osaka City University and Prof. Ito of Osaka Prefecture University, Japan. The experiment done by Prof. Ito and co-workers was conducted in a research project of the Kansai Electric Power Co., Kyushu Electric Power Co., Shikoku Electric Power Co., Hokkaido Electric Power Co., Japan Atomic Power Co. and Mitsubishi Heavy Industries, Ltd. The seismic wave used in the 1/5 reduced-scale tank was provided by staff members related to the project. The author sincerely thanks everyone for their cooperation.

13. References

Abramson, H.N. (1966). *The Dynamic Behavior of Liquid in Moving Containers*, NASA. SP-106, NASA, Washington, D.C., USA.

Akiyama, H.; Takahashi, M. & Nomura, S. (1989). Buckling Tests of Steel Cylindrical Shells Subjected to Combined Bending, Shear and Internal Pressure, *Journal of Structural and Construction Engineering*, No. 400, pp. 113-122 (in Japanese).

Akiyama, H. (1997). *Seismic Resistance of Fast Breeder Reactor Components Influenced by Buckling*, Kajima Institute Publishing, Tokyo, Japan.

Amabili, M. (2003). Theory and Experiments for Large-Amplitude Vibrations of Empty and Fluid-Filled Circular Cylindrical Shells with Imperfections, *Journal of Sound and Vibration* Vol. 262, pp. 921-975.

Amabili, M. & Païdoussis, M.P. (2003). Review of Studies on Geometrically Nonlinear Vibrations and Dynamics of Circular Cylindrical Shells and Panels, with and without Fluid-Structure Interaction, *Applied Mechanics Review*, Vol. 56, No. 4, pp. 349-381.

American Nuclear Society (ANS). (2007). *American National Standard External Events in PRA Methodology*, ANSI/ANS-58.21-2007, ANS, La Grange Park, Illinois, USA.

Architectural Institute of Japan (AIJ). (2010). *Design Recommendation for Storage Tanks and Their Supports (Ver. 4)*, AIJ, Tokyo, Japan (in Japanese).

Atomic Energy Society of Japan (AESJ). (2007). *Standard for Procedure of Seismic Probabilistic Safety Assessment for Nuclear Power Plants*, AESJ-SC-P006:2007, AESJ, Tokyo, Japan (in Japanese).

Belytschko, T.; Lin, J. I. & Tsay, C. S. (1984). Explicit Algorithms for the Nonlinear Dynamics of Shells, *Computer Methods in Applied Mechanics and Engineering*, No. 42, pp. 225-251.

Chiba, M.; Yamaki, N. & Tani, J. (1984). Free Vibration of a Clamped-Free Circular Cylindrical Shell Partially Filled with Liquid; Part I: Theoretical Analysis, *Thin-Walled Structures*, Vol. 2, pp. 265-284.

Chiba, M.; Tani, J.; Hashimoto, H. & Sudo, S. (1986). Dynamic Stability of Liquid-Filled Cylindrical Shells under Horizontal Excitation, Part I: Experiment, *Journal of Sound and Vibration*, Vol. 104, No. 2, pp. 301-319.

Chiba, M., (1993). Experimental Studies on a Nonlinear Hydroelastic Vibration of a Clamped Cylindrical Tank Partially Filled with Liquid, *Journal of Pressure Vessel Technology*, Vol. 115, pp. 381-388.

Clough, R. W.; Niwa, A. & Clough, D. P. (1979). Experimental Seismic Study of Cylindrical Tanks, *Journal of the Structural Division*, Vol. 105, No. 12, pp. 2565-2590.

Fischer, D.F. & Rammerstorfer, F.G. (1982). The Stability of Liquid-filled Cylindrical Shells under Dynamic Loading, In: *Buckling of Shells*, Ramm, E. (ed.), pp. 569-597, Springer-Verlag, Berlin.

Fujii, S.; Shibata, H.; Iguchi, M.; Sato, H. & Akino, K. (1969). Observation of Damage of Industrial Firms in Niigata Earthquake, *Proceedings of 4th World Conference on Earthquake Engineering*, Vol. III, J2-1, pp. 1-19, Santiago, Chile.

Fujita, K. (1981). A Seismic Response Analysis of a Cylindrical Liquid Storage Tank, *Bulletin of the JSME*, Vol. 24, No. 192, pp. 1029-1036.

Fujita, K.; Ito, T.; Matsuo, T.; Shimomura, T. & Morishita, M. (1984). Aseismic Study on the Reactor Vessel of a Fast Breeder Reactor, *Nuclear Engineering and Design*, Vol. 83, pp. 47-61.

Fujita, K.; Ito, T. & Wada, H. (1990). Experimental Study on the Dynamic Buckling of Cylindrical Shell Due to Seismic Excitation, *Proceedings of ASME Pressure Vessels and Piping Conference*, Vol. 191, pp. 31-36, Nashville, Tennessee, USA.

Fujita, K.; Ito, T.; Baba, K.; Ochi, M. & Nagata, K. (1992). Dynamic Buckling Analysis for Cylindrical Shell Due to Random Excitation, *Proceedings of 10th World Conference on Earthquake Engineering*, pp. 4981-4987, Madrid, Spain.

Fujita, K. & Saito, A. (2003). Free Vibration and Seismic Response Analysis of a Liquid Storage Thin Cylindrical Shell with Unaxisymmetric Attached Mass and Stiffness, *Proceedings of ASME Pressure Vessels and Piping Conference*, Vol. 466, pp. 243-252.

High Pressure Gas Safety Institute of Japan (KHK). (2003). *Seismic Design Standard for the High Pressure Gas Facilities*, KHK E 012, KHK, Tokyo, Japan (in Japanese).

Hirt, C.W.; Amsden, A.A. & Cook, J.L. (1974). An Arbitrary Lagrangian-Eulerian Computing Method for All Flow Speeds, *Journal of Computational Physics*, Vol. 4, pp. 227-253.

Housner, G.W. (1957). Dynamic Pressures on Accelerated Fluid Containers, *Bulletin of the Seismological Society of America*, Vol. 47, No. 1, pp. 15-35.

Ibrahim, R.A.; Pilipchuk, V.N. & Ikeda, T. (2001). Recent Advances in Liquid Sloshing Dynamics, *Applied Mechanics Review*, Vol. 54, No. 2, pp. 133-199.

Ito, T.; Morita, H.; Hamada, K.; Sugiyama, A.; Kawamoto, Y.; Ogo, H. & Shirai, E. (2003). Investigation on Buckling Behavior of Cylindrical Liquid Storage Tanks under Seismic Excitation (1st Report; Investigation on Elephant Foot Bulge), *Proceedings of ASME Pressure Vessels and Piping Conference*, Vol. 466, pp. 193-201, Cleveland, Ohio, USA.

Jacobsen, L.S. (1949). Impulsive Hydrodynamics of Fluid inside a Cylindrical Tank and Fluid Surrounding a Cylindrical Pier, *Bulletin of Seismological Society of America*, pp. 189-204.

Japan Electric Association (JEA). (2008). *Technical Codes for Aseismic Design of Nuclear Power Plants*, JEAC4601-2008, JEA, Tokyo, Japan (in Japanese).

Japan Society of Civil Engineers (JSCE). (1989). *Dynamic Analysis and Seismic Design –Energy Facilities–*, pp. 190-191, Gihodo Shuppan Co. Ltd., Tokyo, Japan (in Japanese).

Japan Society of Mechanical Engineers (JSME). (2003). Handbook of Vibration and Buckling of Shells, pp. 243-293, Gihodo Shuppan Co. Ltd., Tokyo, Japan (in Japanese).

Kana, D.D. (1979). Seismic Response of Flexible Cylindrical Liquid Storage Tanks, *Nuclear Engineering and Design*, Vol. 52, pp. 185-199.

Kanagawa Prefecture. (2002). *Seismic Design Standard for the High Pressure Gas Plants*, Kanagawa Prefecture, Yokohama, Japan (in Japanese).

Katayama, H.; Kashima, Y.; Fujimoto, S; Kawamura, Y. & Nakamura, H. (1991). Dynamic Buckling Test Using Scale Model for LMFBR Structure, *Proceedings of 11th International Conference on Structural Mechanics in Reactor Technology (SMiRT 11)*, K15/5, pp. 415-420, Tokyo, Japan.

Livermore Software Technology Corporation. (2003). *LS-DYNA Ver.970 Keyword User's Manual*, Livermore Software Technology Corporation, Livermore, California, USA.

Ma, D.C.; Gvildys, J.; Chang, Y.W. & Liu, W.K. (1982). Seismic Behavior of Liquid-Filled Shells, *Nuclear Engineering and Design*, Vol. 70, pp. 437-455.

Maekawa, A.; Suzuki, M. & Fujita, K. (2006). Nonlinear Vibration Response of a Cylindrical Water Storage Tank Caused by Coupling Effect between Beam-Type Vibration and Oval-Type Vibration: Part 1 Vibration Experiment, *Proceedings of ASME Pressure Vessels and Piping Conference*, PVP2006-ICPVT-11-93261, Vancouver, Canada.

Maekawa, A. & Fujita, K. (2007). Occurrence of Nonlinear Oval-Type Vibration under Large Sinusoidal Excitation: Experiment, *Proceedings of ASME Pressure Vessels and Piping Conference*, PVP2007-26461, San Antonio, Texas, USA.

Maekawa, A.; Fujita, K. & Sasaki, T. (2007). Dynamic Buckling Test of Modified 1/10 Scale Model of Cylindrical Water Storage Tank, *Proceedings of ASME Pressure Vessels and Piping Conference*, PVP2007-26466, San Antonio, Texas, USA.

Maekawa, A. & Fujita, K. (2008). Explicit Nonlinear Dynamic Analysis of Cylindrical Water Storage Tanks Concerning Coupled Vibration between Fluid and Structure, *Proceedings of ASME Pressure Vessels and Piping Conference*, PVP2008-61112, Chicago, Illinois, USA.

Maekawa, A. & Fujita, K. (2009). Dynamic Buckling Analysis of Cylindrical Water Storage Tanks: A New Simulation Method Considering Coupled Vibration between Fluid and Structure, *Proceedings of ASME Pressure Vessels and Piping Conference*, PVP2009-77083, Prague, Czech Republic.

Maekawa, A. & Fujita, K. (2010). Dynamic Buckling Simulation of Cylindrical Liquid Storage Tanks Subjected to Seismic Motions, *Proceedings of ASME Pressure Vessels and Piping Conference*, PVP2010-25412, Bellevue, Washington, USA.

Maekawa, A.; Shimizu, Y.; Suzuki, M. & Fujita, K. (2010). Vibration Test of a 1/10 Reduced Scale Model of Cylindrical Water Storage Tank, *Journal of Pressure Vessel Technology*, Vol. 132, No. 5, pp. 051801-1-051801-13.

Maekawa, A.; Takahashi, T. & Fujita, K. (2011). Seismic Safety Margin of Cylindrical Liquid Storage Tanks in Nuclear Power Plants, *Proceedings of ASME Pressure Vessels and Piping Conference*, PVP2011-57241, Baltimore, Maryland, USA.

National Aeronautics and Space Administration (NASA). (1968). *Buckling of Thin-Walled Circular Cylinders*, NASA Space Vehicle Design Criteria, NASA SP-8007, NASA, Washington, D.C., USA.

Niwa, A. & Clough, R.W. (1982). Buckling of Cylindrical Liquid-Storage Tanks under Earthquake Loading, *Earthquake Engineering and Structural Dynamics*, Vol. 10, pp. 107-122.

Nuclear Safety Commission of Japan (NSC). (2006). *Regulatory Guide for Reviewing Seismic Design of Nuclear Facilities*, NSC, Tokyo, Japan (in Japanese).

Rahnama, M. & Morrow, G. (2000). Performance of Industrial Facilities in the August 17, 1999 Izmit Earthquake, *Proceedings of 12th World Conference on Earthquake Engineering*, 2851, Auckland, New Zealand.

Rammerstofer, F.G.; Scharf, K. & Fischer, F.D. (1990). Storage Tanks under Earthquake Loading, *Applied Mechanics Review*, Vol. 43, No. 11, pp. 261-282.

Shimizu, N. (1990). Advances and Trends in Seismic Design of Cylindrical Liquid Storage Tanks, *JSME International Journal*, Series III, Vol. 33, No. 2, pp. 111-124.

Shin, C.F. & Babcock, C.D. (1980). Scale Model Buckling Tests of a Fluid Filled Tank under Harmonic Excitation, *Proceedings of ASME Pressure Vessels and Piping Conference*, 80-C2/PVP-66, pp. 1-7.

Suzuki, K. (2008). Earthquake Damage to Industrial Facilities and Development of Seismic and Vibration Control Technology, *Journal of System Design and Dynamics*, Vol. 2, No. 1, pp. 2-11.

Tazuke, H.; Yamaguchi, S.; Ishida, K.; Sakurai, T.; Akiyama, H. & Chiba, T. (2002). Seismic Proving Test of Equipment and Structures in Thermal Conventional Power Plant, *Journal of Pressure Vessel Technology*, Vol. 124, pp. 133-143.

Timoshenko, S.P. & Woinowsky-Krieger, S. (1959). *Theory and Plate and Shells (2nd Edition)*, McGraw-Hill, New York, New York, USA.

Timoshenko, S.; Gere, J.M. & Young, D. (1964). *Theory of Elastic Stability (2nd Edition)*, McGraw-Hill, , New York, New York, USA.

Toyoda, Y. & Masuko, Y. (1996). Dynamic Buckling Tests of Liquid-Filled Thin Cylindrical Tanks, *Journal of Structural Engineering*, Vol. 42A, pp. 179-188 (in Japanese).

Toyoda, Y.; Matsuura, S. & Masuko, Y. (1997). Development of Computer Program for Dynamic Buckling Analysis of Liquid-Filled Thin Cylindrical Shell and Its Validation by Experiments, *Journal of Structural Engineering*, Vol. 43B, pp. 31-40 (in Japanese).

Veletsos, A.S. & Yang, J.Y. (1976) Dynamic of Fixed-Based Liquid-Storage Tanks, *Proceedings of U.S.-Japan Seminar for Earthquake Engineering Research with Emphasis on Lifeline Systems*, pp. 317-341.

Werner, P.W. & Sandquist, K.J. (1949). On Hydrodynamic Earthquake Effects, *Transactions of American Geophysical Union*, Vol. 30, No. 5, pp. 636-657.

Yamaki, N. (1984). *Elastic Stability of Circular Cylindrical Shells*, North-Holland, Amsterdam, Netherlands.

3

Seismic Reliability Analysis of Cable Stayed Bridges Against First Passage Failure

Rehan Ahmad Khan
Aligarh Muslim University
India

1. Introduction

In many structural applications, the ultimate purpose of using stochastic analysis is to determine the reliability of a structure, which has been designed to withstand the random excitations. The present study is concerned with one of the failures, which are the results of the dynamic response of the stable cable stayed bridges. If Y(t) is the dynamic response (either deflection, strain or stress) of the bridge at a critical point, the bridge may fail upon the occurrence of one of the following cases (Y.K. Lin, 1967):

i.   Y(t) reaches, for the first time, either an upper level A or lower level –B, where A and B are large positive numbers (Fig. 1(a)).

ii. Damage to the structure accumulated as Y(t) fluctuates at small or moderate excursion which are not large enough to cause a failure of the first type, and failure occurs when the accumulated damage reaches a fixed total.

iii. Failure may occur if a response spends too much of its time off range, i.e., it is. outside the set limits for more than a minimum fraction of its lifetime (Fig. 1(b))

The first case, which is called first passage failure (also called first excursion failure), is the objective of this chapter. According to definition, the first passage failure analysis means determining the probability that a prescribed threshold level (displacement, stress or other response level) will be exceeded, for the first time, during a fixed period of time. The first passage failure does not lead to the catastrophic failure of the bridge, but in view of serviceability consideration it is important. Therefore, the purpose of designing the structures against first passage failure is to reduce the probability of such failure, over the expected lifetime of the structure, to an acceptable level. In most random vibration problems, there is a probability close to unity that any given high response threshold level will be exceeded if the structure is excited for a long enough period of time.

Seismic response of cable stayed bridge due to the random ground motion is obtained in this chapter using frequency domain spectral analysis. The ground motion is assumed to be a partially correlated stationary random process. In the response analysis, the quasi-static response of the bridge deck produced due to the support motion is dully considered. The probability of first passage failure for the possibility of future earthquake is then presented for the analysis of various threshold levels (taken as fraction of yield stress) is considered. A parametric study is performed to show the effect of some important parameters such as threshold level, soil conditions, degree of correlation, angle of incidence of earthquake etc. on the reliability of cable stayed bridge against the first passage failure.

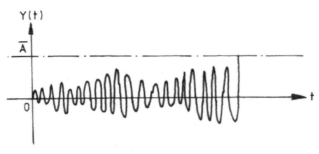

(a) Failure occurs when Y(t) first reaches the level Y(t) = $\overline{A}$.

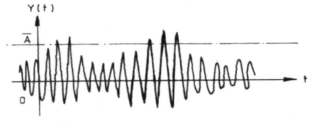

(b) Failure occurs if Y(t) > $\overline{A}$ for more than an acceptable fraction of the total elapse time.

Fig. 1. Possible modes of failure under random excitation.

1.1 Brief review of earlier works

A brief review of the earlier works on different aspects of reliability analysis of structures especially cable supported bridges subjected to seismic forces / dynamic excitations is presented below in order to highlight the need for the present work.

1.1.1 Seismic response of cable supported bridges

The vibration of cable stayed bridges under the earthquake excitation has been a topic of considerable research for many years. The dynamic behaviour of cable stayed bridges, subjected to earthquake ground motion, has been studied by several researchers.

Morris (1974) utilized the lumped mass approach for the linear and non-linear dynamic responses of two dimensional cable stayed bridges due to sinusoidal load applied at a node. It was concluded that four mode solutions were sufficient for both linear and nonlinear solutions. However, a time increment of 10% of the period corresponding to the highest mode gave satisfactory numerical results for all the dynamic analyses carried out.

Abdel Ghaffer and Nazmy (1987) investigated the effects of three dimensionality, multi-support excitation, travelling waves and non-linearity on the seismic behaviuor of cable stayed bridges. Their studies indicated that non-linearities become more important as the span of the bridge increases, and the effect of multiple support seismic excitation is more pronounced for a higher structural redundancy.

Nazmy and Abdel Ghaffar (1991), (1992) studied the effects of non-dispersive travelling seismic wave and ground motion spatial variability on the response of cable stayed bridges considering the cases of synchronous and non-synchronous support motions due to seismic excitations. The responses were obtained by time history analysis. They concluded that depending on the dynamic properties of the local soils at the supports, as well as the soils at the surrounding bridge site, the travelling seismic wave effect should be considered in the seismic analysis of these bridges. Also, there is multi-modal contribution from several modes of vibration to the total response of the bridge, for both displacement and member forces.

Hong Hao (1998) analyzed the effects of various bridge and ground motion parameters on the required seating lengths for bridge deck to prevent the pull-off and -drop collapse using random vibration method. He analyzed two span bridge model with different span lengths and vibration frequencies and subjected to various spatially varying ground excitations. Ground motions with different intensities, different cross correlations and different site conditions were considered in the analysis.

Soyluk and Dumanoglu (2000) carried out asynchronous and stochastic dynamic analysis of a cable stayed bridge (Jindo Bridge) using finite element method. In the asynchronous dynamic analysis, various wave velocities were used for the traveling ground motion. They found that response in the deck obtained from asynchronous dynamic analysis are much higher than the response obtained by the stochastic analysis. Further, shear wave velocity of ground motion greatly influences the response of the Jindo cable stayed bridge.

1.1.2 Reliability analsysis of structures

Lin (1967), Bolotin (1965) and Crandall, Mark (1963), Abbas and Manohar (2005a, b; 2007) discussed various models of structural failure under random dynamic loading and classified them into the following types: (i) failure due to the first excursion of the response beyond a safe level; (ii) failure due to the response remaining above a safe level for too long a duration and (iii) failure due to the accumulation of damage.

All these types of failure are associated with the reliability estimate of Structures subjected to dynamic loading like earthquake. A number of investigations are reported in the literature on the reliability analysis of Structures against the aforementioned types of failure. Relatively recent ones and a few important old ones are reviewed here.

Konishi (1969) studied the safety and reliability of suspension bridges under wind and earthquake actions. He treated the suspension safety of the structure under random ground motion from the standpoint of threshold crossing of a specified barrier.

Vanmarcke (1975) dealt with the problem of the probability of first-passage beyond a threshold value by a time dependent random process. Barrier was classified into three distinct categories: a single barrier, a double barrier, and a barrier defined for the envelope of the random process. The assumption that barrier crossings are independent, so that they constitute a Poisson process, is nearly true for high barrier levels. But for relatively low barrier levels, the Poisson process assumption is in error, as it does not account for the clustering effect in case of a narrow-band process while for a wide-band process, the actual time spent in the unsafe domain is not considered. Vanmarcke suggested improvements in the Poisson process assumption to allow for the above effects.

Chern (1976) dealt with the reliability of a bilinear hysteretic system, subjected to a random earthquake motion, considering first excursion beyond a specified barrier and low-cycle fatigue.

Solomon and Spanos (1982) studied the structural reliability under a non-stationary seismic excitation, based on the first excursion beyond a specified barrier by the absolute value and envelope of the response process.

Schueller and stir (1987) reviewed various methods to calculate the failure probabilities of structural component or system in the light of their accuracy and computational efficiency.

They analyzed that for problems of higher dimensions, approximation techniques utilizing linearization like FOSM introduced considerable error. In view of these difficulties, they provided an alternate method to calculate the failure probabilities, which combines the advantages of both the importance sampling technique and the design point calculation. Mebarki et al. (1990) presented a new method for reliability assessment, called Hypercone method based on the principles of level-2 methods of reliability analysis. The main aim of the method was to evaluate reliability index 'beta' and to deduce the values of the probability of failure by considering the whole geometry of the failure domain. The restrictions and the practical application of the method were discussed. The results reported showed that the value of probability of failure deduced from the Hasofer-Lind index beta, assuming linearity of the limit state surface, and from Monte Carlo Simulations are in accordance with those deduce from the Hypercone method.

Zhu (1993) reviewed several methods for computing structural system reliability. A discretization or cell technique for determining the failure probabilities of structural system is proposed. The gaussian numerical integration method is introduced to improve its computational accuracy and can be applied to gaussian or non-gaussian variables with linear or non-linear safety margin. Harichandran et al. (1996) studied the stationary and transient response analyses of the Golden Gate suspension bridge, and the New River Gorge and Cold Spring Canyon deck arch bridges subjected to spatially varying earthquake ground motion (SVEGM). They found that transient lateral displacements of the suspension bridge center span significantly overshoot the corresponding stationary displacements for the filtered Kanai-Tajami excitation power spectrum; this spectrum may therefore be unsuitable for analyzing very flexible structures.

2. Assumptions

The following assumptions are made for the First Passage Failure analysis:

i.   It is assumed that the response process is stationary Gaussian with zero mean value.

ii.   As in spectral analysis method, the evaluated response is always positive, therefore the single barrier level (called type B barrier, according to Crandall et al.1966) is used.

iii.   It is assumed that the threshold level crossing occur independently according to a Poisson process.

iv.   The structure is assumed to be linear and lightly damped.

v.   The bridge deck (girder) and the towers are assumed to be axially rigid.

vi.   The bridge deck, assumed as continues beam, does not transmit any moment to the towers through the girder-tower connection.

vii.   Cables are assumed to be straight under high initial tensions due to the dead load and well suited to support negative force increment during vibration without losing its straight configuration.

viii. An appropriate portion of the cable is included (in addition to deck mass) in the dynamic analysis of the bridge deck and is assumed to be uniformly distributed over the idealized deck.

ix.   Beam-column effect, in the stiffness formulation of the beam is considered for the constant axial force in the beam and its fluctuating tension in the cable is ignored. Further, cable dynamics is ignored for the bridge deck vibration, i.e., the tension fluctuations in the cables are assumed to be quasi-static, and do not introduce any nonlinearity in the system.

3. Seismic excitation

The seismic excitation is considered as a three component stationary random process. The earthquake ground motion is assumed as stationary random although in many cases it is assumed as a uniformly modulated non-stationary process. The response analysis remains the same for both cases. The response derived by assuming the process to be stationary can be multiplied by an envelope function to take care of the non-stationary. The components of the ground motion along an arbitrary set of orthogonal directions will be usually statistically correlated. However, as observed by Penzien and Watable (1975), the three components of ground motion along a set of principal axes are uncorrelated. These components, directed along the principal axes, are usually such that the major principal axis (u) is directed towards the expected epicenter, the moderate principal axis (w) is directed perpendicular to it (horizontally) and the minor principal axis (v) is directed vertically as shown in Fig.2. Nigam and Naranayan (1995) highlight the critical orientation of the alpha angle between the two sets of axes. Der Kiureghian (1996) developed the model for the coherency function describing spatial variability of earthquake ground motions. The model consists of three components characterizing three distinct effects of spatial variability, namely, the incoherence effect that arises from scattering of waves in the heterogeneous medium of the ground and their differential super positioning when arriving from an extended source, the wave-passage effect that arises from difference in the arrival times of waves at different stations, and the site-response effect that arises from difference in the local soil conditions at different stations. Abbas and Manohar (2002) developed a critical earthquake excitation models with emphasis on spatial variability characteristics of ground motion. In this study, the three components of the ground motion are assumed to be directed along the principal axes of the bridge x, y, z or shifted with an angle α. Each component is assumed to be a stationary random and partially correlated process with zero mean characterized by a psdf. The psdf of the ground acceleration is defined by Clough and Penzien (1975) as

$$S_{\ddot{f}_g \ddot{f}_g} = \left| H_1(i\omega) \right|^2 \left| H_2(i\omega) \right|^2 S_o \tag{1}$$

in which S_o is the spectrum of the white noise bedrock acceleration; $\left| H_1(i\omega) \right|^2$ and $\left| H_2(i\omega) \right|^2$ are the transfer functions of the first and the second filters representing the dynamic characteristic of the soil layers above the bedrock, where

$$\left| H_1(i\omega) \right|^2 = \frac{1 + \left(2\xi_g \omega / \omega_g \right)^2}{\left[1 - \left(\omega / \omega_g \right)^2 \right]^2 + \left(2\xi_g \omega / \omega_g \right)^2} \tag{2}$$

$$\left| H_2(i\omega) \right|^2 = \frac{\left(\omega / \omega_f \right)^4}{\left[1 - \left(\omega / \omega_g \right)^2 \right]^2 + \left(2\xi_g \omega / \omega_g \right)^2} \tag{3}$$

in which ω_g, ξ_g are the resonant frequency and damping ratio of the first filter, and ω_f, ξ_f are those of the second filter. The clough and Penzien is a double filter for spectral density of the ground acceleration for which the corresponding displacement psdf does not become unrealistic. In case of Kanai-Tajimi spectrum, although the psdf of acceleration is simpler but it has problem that the corresponding displacement psdf become undefined at zero frequency.

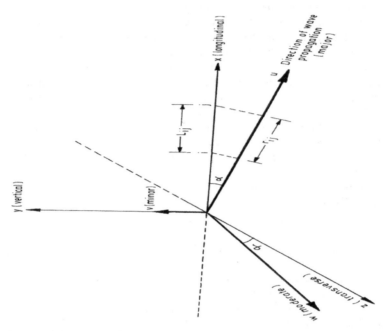

Fig. 2. Principal directions of the bridge(x,y,z) and the ground motion(u,v,w)

The cross spectrum between the random ground motions $\ddot{f}_{g_i}$ and $\ddot{f}_{g_j}$ at two stations i and j is described by that given by Hindy and Novak (1980) as

$$S_{\ddot{f}_{g_i}\ddot{f}_{g_h}}\left(r_{ij},\omega\right) = S_{\ddot{f}_g\ddot{f}_g}\left(\omega\right)\rho_{ij}\left(\omega\right) \tag{4}$$

in which $S_{\ddot{f}_g\ddot{f}_g}\left(\omega\right)$ local spectrum of ground acceleration as given in Eqn.(1) which is assumed to be the same for all supports and $\rho_{ij}\left(\omega\right)$ is the cross correlation function (coherence function) of the ground motion between two excitation points i, j and is represented by

$$\rho_{ij}\left(\omega\right) = \exp\left[-c\left(\frac{r_{ij}\omega}{2\pi V_s}\right)\right] \tag{5}$$

in which r_{ij} is the separation distance between stations i and j measured in the direction of wave propagation; c is a constant depending upon the distance from the epicenter and the inhomogeneity of the medium; V_s is the shear wave velocity of the soil; and ω is the frequency (rad/sec) of the ground motion.

For one sided spectrum it is well known that

$$\sigma_{\ddot{f}_g}^{2} = S_0 \left[\int_0^{\alpha} |H_1(i\omega)|^2 |H_2(i\omega)|^2 \, d\omega \right] \tag{6}$$

$\sigma_{\ddot{f}_g}^{2}$ is the variance of ground acceleration. The Modified Mercalli intensity of an earthquake Is is a measures of the strength of the shaking at a particular location and will vary with distance, substrate conditions and other factors. The empirical relation between the standard deviation of peak ground acceleration and earthquake intensity Is is given as

$$\sigma_{\ddot{f}_g}^{2} = 10^{(I_s/3-0.5)} / K^* \tag{7}$$

where K* is a peak factor given as

$$K^* = K' + \frac{0.5772}{K'} \; ; \; K' = \sqrt{2\ln(N_0 T)} \tag{8}$$

The empirical relation between the magnitude of earthquake and intensity of earthquake is given as

$$M = 1.3 + 0.6 \, I \tag{9}$$

In Eqn.(9), N_0 is the mean rate of zero crossing and is given by

$$N_0 = \frac{1}{2\pi} \sqrt{\int_0^{\alpha} \omega^2 S_{\ddot{f}_g}(\omega) d\omega / \int_0^{\alpha} S_{\ddot{f}_g}(\omega) d\omega} \tag{10}$$

By defining the filter characteristics ω_g, ξ_g, ω_f, ξ_f and specifying a standard deviation of the ground acceleration $\sigma_{\ddot{f}_g}$, the psdf of the ground acceleration can be completely defined. The psdfs $S_{f_g f_g}(\omega)$ and $S_{\dot{f}_g \dot{f}_g}(\omega)$ of the ground displacement and velocity respectively are related to $S_{\ddot{f}_g \ddot{f}_g}(\omega)$ by

$$S_{f_g f_g}(\omega) = S_{\ddot{f}_g \ddot{f}_g}(\omega) / \omega^4$$

$$S_{\dot{f}_g \dot{f}_g}(\omega) = S_{\ddot{f}_g \ddot{f}_g}(\omega) / \omega^2 \tag{11}$$

The ground motion is represented along the three principal directions (u,v,w) by defining ratio R_u, R_v and R_w along them such that

$$\ddot{u}_g(t) = R_u \ddot{f}_g(t) \; ; \; \ddot{v}_g(t) = R_v \ddot{f}_g(t) \; ; \; \ddot{w}_g(t) = R_w \ddot{f}_g(t) \tag{12}$$

and the psdfs of the ground acceleration in the principal directions of the ground motion (u,v,w) can be defined as

$$S_{\ddot{u}_g \ddot{u}_g}(\omega) = R_u^2 S_{\ddot{f}_g \ddot{f}_g}(\omega) \; ; \; S_{\ddot{v}_g \ddot{v}_g}(\omega) = R_v^2 S_{\ddot{f}_g \ddot{f}_g}(\omega) \; ; \; S_{\ddot{w}_g \ddot{w}_g}(\omega) = R_w^2 S_{\ddot{f}_g \ddot{f}_g}(\omega) \tag{13}$$

so that

$$\sigma_{\ddot{u}_g}^{\ 2} = R_u^{\ 2}\sigma_{\ddot{f}_g}^{\ 2} \quad ; \quad \sigma_{\ddot{v}_g}^{\ 2} = R_v^{\ 2}\sigma_{\ddot{f}_g}^{\ 2} \quad ; \quad \sigma_{\ddot{w}_g}^{\ 2} = R_w^{\ 2}\sigma_{\ddot{f}_g}^{\ 2} \tag{14}$$

When the angle of incidence of the ground motion with respect to the principal direction of the bridge is defined as α , the ground motions along the principal directions of the bridge (x,y,z) are defined as

$$\ddot{x}_g(t) = \ddot{u}_g(t)\cos\alpha - \ddot{w}_g(t)\sin\alpha$$
$$\ddot{z}_g(t) = \ddot{u}_g(t)\sin\alpha - \ddot{w}_g(t)\cos\alpha \tag{15}$$
$$\ddot{y}_g(t) = \ddot{v}_g(t)$$

The psdfs of the ground accelerations along x,y,z can be written as

$$S_{\ddot{x}_g\ddot{x}_g} = \cos^2\alpha\ S_{\ddot{u}_g\ddot{u}_g} + \sin^2\alpha\ S_{\ddot{w}_g\ddot{w}_g} = R^2_{\ x}S_{\ddot{f}_g\ddot{f}_g}$$

$$S_{\ddot{z}_g\ddot{z}_g} = \sin^2\alpha\ S_{\ddot{u}_g\ddot{u}_g} + \cos^2\alpha\ S_{\ddot{w}_g\ddot{w}_g} = R^2_{\ z}S_{\ddot{f}_g\ddot{f}_g}$$

$$S_{\ddot{y}_g\ddot{y}_g} = S_{\ddot{v}_g\ddot{v}_g} = R_y^{\ 2}S_{\ddot{f}_g\ddot{f}_g} \tag{16}$$

where R_x, R_y ,and R_z are the ratios of the ground motion along the principal axes of the bridge and given as

$$R_x^{\ 2} = R_u^{\ 2}\cos^2\alpha + R_w^{\ 2}\sin^2\alpha$$
$$R_z^{\ 2} = R_u^{\ 2}\sin^2\alpha + R_w^{\ 2}\cos^2\alpha \tag{17}$$
$$R_y^{\ 2} = R_v^{\ 2}$$

and R_u , R_v , and R_w are ratios of the ground motion along the principal directions of the ground motion (u,v,w) as shown in Fig.2.

4. Distribution function of magnitude of earthquake

Two types of distribution functions of the magnitude of earthquake are considered in the study.

4.1 Exponential distribution
This type of probability distribution function of the magnitude of earthquake is based on the Gutenberg – Ritcher Recurrence law (Kagan, 2002a)

$$\log \lambda_m = a - b\ m \tag{18a}$$

$$\lambda_m = 10^{a-b\ m} = \exp(\alpha - \beta\ m) \tag{18b}$$

where λ_m is the mean annual rate of exceedence of magnitude m ; 10^a is the mean yearly number of earthquakes greater than or equal to zero,; and b describes the relative likelihood of large or small earthquakes. Eqn.(18b) implies that the magnitudes are exponentially distributed. Based on Eqn. (18b), the probability density function (PDF) is given by

$$P_M(m) = \beta e^{-\beta(m-m_0)} \tag{19}$$

where, $\beta = 2.303b$, and m_0 is the lower threshold magnitude of earthquake, earthquakes smaller than which are eliminated, and m is the magnitude of earthquake.
The cumulative distribution function (CDF) of magnitude of earthquake for exponential distribution is given by the following expression

$$F_M(m) = \{1-\exp(-\beta(m-m_0))\} \tag{20}$$

4.2 Gumbel type-I distribution
The cumulative distribution function of magnitude of earthquake for Gumbel type-I distribution is given by the following expression

$$F(m) = \text{Exp}(-\exp - \alpha\,(m - u)) \tag{21}$$

where α and u are the parameters for Gumbel Type-1 distribution given by

$$\bar{M} = u + 0.5772/\alpha \tag{22a}$$

$$\sigma_m^2 = \pi^2/6\,\alpha^2 \tag{22b}$$

in which $\bar{M}$ and σ_m are the mean and standard deviation of the magnitudes of earthquake respectively.

5. Theoretical analysis

5.1 Free vibration analysis of cable stayed bridge deck (girder)
The bridge deck, as shown in Figs.3(a) is idealized as a continuous beam over the outer abutments and the interior towers, and the effect of cable is taken as vertical springs at the points of intersections between the cables and the bridge deck shown in Fig.3(b). Further, the effect of the spring stiffness is taken as an additional vertical stiffness to the flexural stiffness of the bridge.
Referring to Fig.3(c) and Fig.3(d), the fluctuations of tension in the cable at any instant of time (t) can be written as

$$h_i(t) = K_i V(x_i,t)\sin\psi_i + K_i\Delta_j(t)\cos\psi_i \tag{23}$$

where, $K_i = E_c A_i / L_i$ is the stiffness of the ith cable; $V(x_i,t)$ is the displacement of the girder at time t at the joint of the ith cable with the girder; $\Delta_j(t)$ is the horizontal sway of the tower at the ith tower cable joint connecting the ith cable; ψ_i is the angle of inclination of the ith cable to the horizontal (measured clockwise from the cable to the horizontal line as shown in Fig.3.3(c); A_i, L_i are the cross-sectional area and the length of the ith cable respectively and E_c is the equivalent modulus of elasticity of the straight cable under dead loads.

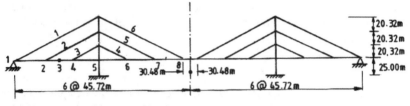

(a) Harp type cable stayed bridge considered for parametric study(Bridge-I)

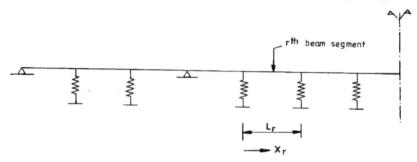

(b) Idealization of the bridge deck

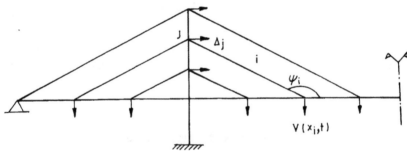

(c)Displacement due to the fluctuation of the ith cable

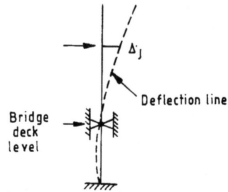

(d)Main system and the displacement of the tower

Fig. 3. (a, b, c, d) Problem identification

Following Eqn.(23), the changes in tensions in the array of cables can be put in the following matrix form

$$\{h\}_{Ncx1} = (A)_{NcxNd} \{V\}_{Ndx1} + (B)_{Ncx1}\{\Delta\}_{Nt\,x1} \tag{24}$$

Where Nc is the number of cable (or pair of cable in case of two-phase cable stayed bridge); Nd is the number of unrestrained vertical degrees of freedom of the girder at the cable-girder joints; Nt is the number of horizontal tower degrees of freedom at the cable-tower joints; $\{V\}$ and $\{\Delta\}$ are the girder and the tower displacement vectors; $\{h\}$ is the vector of incremental cable tensions; (A) and (B) are the matrices which are formed by proper positioning of the elements $K_i \sin\psi_i$ and $K_i \cos\psi_i$ respectively.

The deflections of the tower at the cable joints can be obtained by assuming that the tower behaves like a vertical beam fixed at the bottom end and constrained horizontally at the level of the bridge deck and subjected to the transverse forces h_i (t) $\cos\psi_i$ (i=1, Nc) at the cable tower joints as shown in Fig.3(b) and are given by

$$\{\Delta\} = (C)\{h\} \tag{25}$$

where the elements of the matrix (C) can be easily obtained from the deflection equation of the tower (vertical beam) subjected to concentrated load, as mentioned above, using standard structural analysis procedure. Eliminating $\{\Delta\}$ from Eqns.(24) and (25), the relation between the vectors of incremental cable tensions and girder deflections may be written as

$$\{h\} = ((I)-(B)\,(C))^{-1}\,(A)\{V\} \tag{26}$$

where (I) is the unit matrix of order Nc.

Premultiplying both sides of Eqn. (26) by a diagonal matrix (D) of order Nc, where the diagonals consists of the terms of $\sin\psi_i$ (i= 1 to Nc), Eqn. (26) can be written as

$$\{h_{v.b}\} = (K_{c,v})\,\{V\} \tag{27}$$

where $\{h_{v,b}\}$ is the vector containing the vertical components of incremental cable tensions, and $(K_{c,v}) = ((D)(I)-(B)(C))^{-1}(A)$ is the stiffness matrix of the bridge contributed by the cables in vertical vibration.

The equation of motion for the relative vertical vibration $Y(x_r,t)$ of the beam segment r of the idealized deck with constant axial force N_r, neglecting the shear deformation and rotary moment of inertia is given by

$$E_d I_r \frac{\partial^4 Y}{\partial X_r^4} + N_r \frac{\partial^2 Y(x_r,t)}{\partial X_r^2} + C_r \frac{\partial Y(x_r,t)}{\partial t} + \frac{\overline{W}_r}{g}\frac{\partial^2 Y(x_r,t)}{\partial t^2} = P(x_r,t) \text{ for r=1,2,3 ----. N_b} \tag{28}$$

in which E_d and I_r are the modulus of elasticity of the bridge deck and vertical moment of inertia of the beam segment r of the deck respectively.

$P(x_r,t)$ is defined as the load induced due to seismic excitations at different support degree of freedom and is given by

$$P(X_r,t) = -\frac{\overline{W}_r}{g}\sum_{j=1}^{8} g_{jr}(x_r)\ddot{f}_j(t) \tag{29}$$

where $\ddot{f}_j(t)$, j=1,2, --------, 8 are the accelerations at the different support degrees of freedom and $g_{jr}(x_r)$ is the vertical displacement of the rth segment of the bridge deck due to unit displacement at the jth degree of the supports. In Eqns. (28) and (29), it is assumed that for lightly damped system, the effect of damping term associated with quasi-static movement of the supports is negligible (Clough and Penzien, 1993).

The expression for n^{th} mode shape (undamped) for vertical vibration of the rth segment of the bridge deck is given by (Chatterjee, 1992)

$$\phi_n(x_r) = A_{nr}\cos\beta_{nr}x_r + B_{nr}\sin\beta_{nr}x_r + C_{nr}\cosh\gamma_{nr}x_r + D_{nr}\sinh\gamma_{nr}x_r \tag{30}$$

where A_{nr}, B_{nr}, C_{nr} and D_{nr} are the integration constants expressed in terms of nth natural frequency of vertical vibration ω_{bn} and

$$\beta_{nr} = \sqrt{\frac{N_r(Z_{nr}+1)}{2E_d I_r}} \quad ; \quad \gamma_{nr} = \sqrt{\frac{N_r(Z_{nr}-1)}{2E_d I_r}}$$

where

$$Z_{nr} = \sqrt{1 + \frac{4E_d I_r \overline{W}_r / g\omega_{bn}^2}{N_r^2}}$$

in which the suffix r is used to mean the rth segment of the beam. The origin for the rth segment is fixed at the left end as shown in Fig.3.3(c).Utilizing Eqn.(30), a relation between end displacements (vertical deflection and slope) and end forces(shear forces and bending moments) for the rth segment may be written as

$$\{F\}_r = (K)_r\{x_r\} \tag{31}$$

where $\{F\}_r$ and $\{x_r\}$ are the end forces and end displacement vectors and $(K)_r$ is the flexural dynamic stiffness matrix of the rth beam segment. The integration constants A_{nr}, B_{nr}, etc. are related to the end displacements as

$$\{C\}_r = (T)_r\{x_r\} \tag{32}$$

where $\{C\}_r$ is the vector of integration constants containing A_{nr} etc., and $(T)_r$ is the matrix integration constants. The sign conventions used in the dynamic stiffness formulation are shown in Fig.4. The explicit expressions for the elements of $(K)_r$ and $(T)_r$ are given by Chatterjee (1992). Assembling the stiffness $(K)_r$ for each element (r) and adding the vertical stiffness due to cables $(K_{C,V})$, the overall stiffness of the bridge (K) is obtained. The condition for the free vibration of the bridge deck may then be written as

$$(K)\{U\} = \{0\} \tag{33}$$

where $\{U\}$ is the unknown end displacement vector for the beam corresponding to the dynamic degrees of freedom Fig.4. Using Eqn.(33) leads to

$$\det (K) = 0 \tag{34}$$

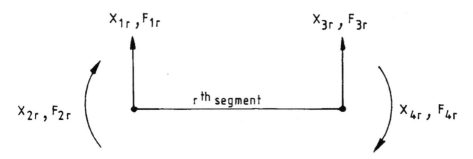

Fig. 4. Sign conventions used in the dynamic stiffness formulation.

Using Regula Falsi approach, the natural frequencies for the system are determined from the solution of Eqn.(33). Once the natural frequencies are obtained, mode shapes can be known through the use of Eqns.(32 & 33).

5.2 Modal transformation

The modal transformation of the relative vertical displacement $y(x_r,t)$ for any point in the rth deck segment is given as

$$y(x_r,t) = \sum_{n=1}^{\alpha} \varphi_n(x_r) q_n(t) \quad r = 1,2,----,N_b \tag{35}$$

in which $\mathfrak{c}_n(x_r)$ is the nth mode shape of the rth beam segment of the bridge deck and $q_n(t)$ is the nth generalized coordinate. Substituting Eqn.(35) into Eqn.(28), multiplying by $\mathfrak{c}_m(x_r)$, integrating w.r.t. L_r and using the orthogonality of the mode shapes leads to

$$\ddot{q}_n(t) + 2\xi_n \omega_n \dot{q}_n(t) + \omega_n^2 q_n(t) = \bar{P}_n(t) \quad n = 1, ----, M \tag{36}$$

in which ζ_n and ω_n are the damping ratio and the natural frequency of the nth vertical mode; M is the number of modes considered and $\bar{P}_n(t)$ is the generalized force given as

$$\bar{P}_n(t) = \sum_{j=1}^{8} R_{jn}(x_r) \ddot{f}_j(t) \tag{37}$$

where R_{jn} is the modal participation factor given by

$$R_{jn} = -\frac{\displaystyle\sum_{r=1}^{N_b} \frac{\bar{W}_r}{8} \int_0^{L_r} g_{jr}(x_r) \phi_n(x_r) dx_r}{\displaystyle\sum_{r=1}^{N_b} \frac{\bar{W}_r}{8} \int_0^{L_r} \phi_n^2(x_r) dx_r} \tag{38}$$

in which $g_{jr}(x_r)$, the quasi static function, is the vertical displacement of the rth beam segment of the bridge deck due to unit displacement given in the jth direction of support movement.

Eqn.(37) can be put in the following matrix form

$$\overline{P}_n(t) = [G_n]\{f\} \tag{39}$$

in which

$$[G_n] = \{G_{1n} \text{ --------} G_{8n}\} \text{ and } \quad \{f\}^T = \left\{ \ddot{f}_1(t) \text{------} \ddot{f}_8(t) \right\}$$

where $[G_n]$ is the generalized force coefficients for the nth mode and can be obtained by Eqns.(36 & 37) i.e. $G_{1n} = R_{1n}, \text{-----}, G_{8n} = R_{8n}.$

5.3 Spectral analysis
5.3.1 Evaluation of the relative displacement
Applying the principles of modal analysis, the nth generalized coordinate in frequency domain can be written as

$$q_n(\omega) = H_n(\omega)\overline{P}_n(\omega) \tag{40}$$

in which $H_n(\omega)$ is the nth modal frequency response function given by

$$H_n(\omega) = \left[\left(\omega_n^2 - \omega^2 \right) + i \left(2\xi_n \omega_n \omega \right) \right]^{-1} \tag{41}$$

Similarly, the mth generalized coordinate can be written as

$$q_m(\omega) = H_m(\omega)\overline{P}_m(\omega) \tag{42}$$

The cross power spectral density function between the two generalized coordinate $q_n(\omega)$ and $q_m(\omega)$ is given by

$$S_{q_n q_m}(\omega) = H_n(\omega)H_m(\omega)S_{\overline{P}_n \overline{P}_m} \tag{43}$$

$H_n(\omega)$ denotes the complex conjugate of the $H_n(\omega)$ and $S_{\overline{P}_n \overline{P}_m}$ can be written in the matrix form as

$$S_{\overline{P}_n \overline{P}_m} = [G_n]\left[S_{ff} \right][G_m]^T \tag{44}$$

$[S_{ff}]$ is the psdf matrix for the ground motion inputs (of size 8x 8) which are the support accelerations i.e. $\ddot{f}_1(t)$, $\ddot{f}_2(t)$, $\ddot{f}_3$, $\ddot{f}_4$, $\ddot{f}_5$, $\ddot{f}_6$, $\ddot{f}_7$, $\ddot{f}_8$.

Any element of the matrix $[S_{ff}]$ may be written in terms of the psdf of ground acceleration $S_{\ddot{f}_g \ddot{f}_g}(\omega)$ by using the coherence function Eqn.(5), the ratio between the three components of the ground motion $(R_u:R_v:R_w)$ and Eqns.(16 & 17) as explained earlier. For example (1,4) and (5,8) can be written in the form

$$S_{\dot{f}_1 \dot{f}_4} = \rho_{14}(\omega) R_y{}^2 S_{\dot{f}_g \dot{f}_g} \tag{45}$$

$$S_{\dot{f}_5 \dot{f}_8} = \rho_{58}(\omega) R_x{}^2 S_{\dot{f}_g \dot{f}_g} \tag{46}$$

Using the expression given in Eqn.(42), the elements of the matrix $[S_{qq}]$ may be formed which has the dimension of M x M. Since the relative displacement $y(x_r,t)$ is given by

$$y(x_r,t) = \sum_{n=1}^{M} \phi_n(x_r) q_n(t) = \left[\phi(x_r)\right]_{(1\times m)} \{q\}_{(1\times m)} \tag{47}$$

The psdf of the response $y(x_r,t)$ is given by

$$S_{yy}(x_r,\omega) = \left[\phi(x_r)\right]\left[S_{qq}\right]\left[\phi(x_r)\right]^T \tag{48}$$

5.3.2 Evaluation of the quasi-static displacement
The quasi-static component of the vertical displacement at any point in the rth deck segment at time (t) is given as

$$g(x_r,t) = \sum_{j=1}^{8} g_{jr}(x_r) f_j(t) = [Q]\{\bar{f}\} \tag{49}$$

where

$$[Q] = \{ g_{1r}(x_r)\ g_{2r}(x_r)\ \text{-----------}\ g_{8r}(x_r)\}; \quad \left\{\bar{f}\right\}^T = \{f_1(t)\ f_2(t)\ \text{---------}\ f_8(t)\}$$

$g_{jr}(x_r)$ is the vertical displacement at any point in the rth beam segment of the bridge deck due to unit movement of the jth support d.o.f. The psdf of the quasi-static displacement at any point in the rth deck segment is given by

$$S_{gg}(x_r,\omega) = [Q]\left[S_{\bar{f}\bar{f}}\right][Q]^T \tag{50}$$

where $\left[S_{\bar{f}\bar{f}}\right]$ is the psdf matrix for the ground displacements at the support d.o.f.s and can be obtained in terms of the psdf of ground acceleration $S_{\dot{f}_g \dot{f}_g}$ with the help of both the coherence function, the ratio between the three components of ground motion and Eqns.(11, 16 & 17).

5.3.3 Evaluation of the total displacement
The total displacement at any point of the rth beam segment of the bridge deck at any time (t) can be written as

$$Y(x_r,t) = y(x_r,t) + g(x_r,t) \tag{51}$$

The psdf of the total vertical displacement can be expressed as

$$S_{YY}(x_r,\omega) = S_{yy}(x_r,\omega) + S_{gg}(x_r,\omega) + S_{yg}(x_r,\omega) + S_{gy}(x_r,\omega) \tag{52}$$

$S_{yg}(x_r,\omega)$ and $S_{gy}(x_r,\omega)$ are the cross power spectral density functions between relative and the quasi-static displacements. Using Eqns.(--------), the expression for $S_{yg}(x_r,\omega)$ can be obtained as

$$S_{yg}(x_r,\omega) = \underset{(1\times M)}{[\phi(x_r)]} diag \underset{M\times M}{[Hn(\omega)]} \underset{M\times 8}{[G]} \underset{8\times 8}{[S_{\bar{f}\bar{f}}]} \{g_{jr}(x_r)\}_{8\times 1} \tag{53}$$

$$n=1, 2, ----, M; \quad j = 1,2, -----,8; \quad r =1, 2, ------,N_b$$

where $[S_{\bar{f}\bar{f}}]$ is the cross power spectral density matrix of the random vectors $\{\bar{f}\}$ and $\{\bar{f}\}$ i.e., the support accelerations and displacements. $S_{gy}(x_r,\omega)$ is the complex conjugate of $S_{yg}(x_r,\omega)$.

5.3.4 Evaluation of the bending moment

Using Eqns.(45 & 47) and differentiating the expression for $Y(x_r, t)$ twice with respect to x, the following expression for the bending moment can be obtained as

$$E_d I_r \frac{\partial^2 y}{\partial x^2} = \sum_{n=1}^{M} E_d I_r \frac{d^2 \phi(x_r)}{dx^2} q_n(t) + \sum_{j=1}^{8} E_d I_r \frac{d^2 g_{jr}(x_r)}{dx^2} f_j(t) \tag{54}$$

Similar expressions can be obtained for the psdf of the bending moment at any point in the rth beam segment of the bridge deck as those derived for the total displacement by replacing $\phi(x_r)$ and $g_{jr}(x_r)$ by $E_d I_r d^2\phi(x_r)/dx^2$ and $E_d I_r d^2 g_{jr}(x_r)/dx^2$ respectively. $E_d I_r d^2 g_{jr}(x_r)/dx^2$ is obtained from the quasi-static analysis of the entire bridge using the stiffness approach as mentioned before.

5.4 Statistical parameters of response

For studying the statistical properties of the response process, the first few moments of the response power spectral density function are needed. The jth moment of the PSDF function may be defined as follows:

$$\lambda_j = \int_0^\alpha w^j S_{YY}(w) dw \quad \text{where } j = 0, 1, 2,---- \tag{55}$$

The zeroth and second moments may be recognized as the variances of the response and the first time derivative of the response respectively.

$$\lambda_0 = \int_0^\alpha S_{YY}(w) dw = \sigma_{YY}^2 \tag{56}$$

$$\lambda_2 = \int_0^\alpha w^2 S_{YY}(w) dw = \sigma_{\dot{Y}\dot{Y}}^2 \tag{57}$$

The mean rate of zero crossing at positive slopes is by

$$v_0 = \frac{1}{2\pi}\sqrt{\frac{\lambda_2}{\lambda_0}} \tag{58}$$

Another quantity of interest is the dispersion parameter q given by

$$q = \sqrt{1 - \frac{\lambda_1^2}{\lambda_0\lambda_2}} \tag{59}$$

The value of q lies in the range [0,1]. It can shown that q is small for a narrow band process and relatively large for a wide band process. The mean rate of crossing a specified level A at a positive slope by a stationary zero mean Gaussian random process z(t) can be expressed by Lin (1967) as

$$v_a = v_0 e^{\frac{-\psi^2}{2}} \tag{60}$$

where

$$\psi = \frac{A}{\sigma_{YY}}$$

It has been confirmed by theoretical as well as simulation studies that the probability of a stationary response process remaining below a specified barrier level decays approximately exponentially with time as given by the relationship (Coleman,1959 and Crandal et.al., 1966)

$$L(T) = L_o e^{-\alpha T} \tag{61}$$

Where, L_0 is the probability of starting below the threshold, α is the decay rate, and T is the duration of the response process.

At high barrier levels, L_o is practically equal to one, and the decay rate is given by the following expressions for processes with double barrier and one sided barrier respectively (Vanmarcke, 1975; Lin, 1967)

$$\alpha_D = 2v_a \tag{62}$$

$$\alpha_S = v_a \tag{63}$$

In case of relatively low threshold levels, an improved value can be obtained by using the expressions for the probability of starting below the threshold and the decay rate (Vanmarcke, 1975)

$$L_o = 1 - e^{\frac{-\psi^2}{2}} \tag{64}$$

$$\alpha_D{}^* = \alpha_D \frac{1 - e^{-\sqrt{\frac{\pi}{2}}(q\psi)}}{1 - e^{-\psi^2/2}} \tag{65}$$

$$\alpha_S^* = \alpha_S \frac{1 - e^{-\sqrt{2\pi} \, (q\psi)}}{1 - e^{-\psi^2/2}} \tag{66}$$

5.5 Reliability estimation against first passage failure

For an earthquake with given magnitude M, the probability of First Passage Failure, i.e., the probability that the response is larger than a threshold level A, can be determined from the following relationship

$$p\left[z > A | M_r\right] = 1 - L(T) \tag{67}$$

where T is the duration of the response.

If f_M (M) is the probability density functions of earthquake magnitude, the probability of First Passage Failure, provided that an earthquake occurs, can be calculated from (Ang and Tang, 1975)

$$p_E = p[z > A] = \int_{M=4}^{9} p[z > A | M_r] f_M(M) dM \tag{68}$$

If the rate of earthquake occurrence for the seismo – tectonic region considered in the study is a constant and n is the average number of earthquakes per year in the magnitude range of interest for the source region, the probability of atleast one failure due to earthquake in "m" years can be expressed as

$$P_F = 1 - (1 - p_E)^{m.n} \tag{69}$$

6. Numerical study

A double plane symmetrical harp type cable stayed bridges, (Morris, 1974) used as illustrative example is shown in Fig.3.3(a). The structural data of the bridge is shown in Table-1.

In addition, the following data are assumed for the analysis of the problem, $E_c = E_d$; $\xi = 0.02$ for all modes; and the tower – deck inertia ratio, the ratio between three components of the ground motion ($R_u:R_v:R_w$), α, duration of earthquake, β values for exponential distribution of magnitude of earthquake are taken 4, (1.0:1.0:1.0) , 0.0^0, 15 sec and (1.5, 2.303, 2.703) respectively unless mentioned otherwise. Also, the ground motion is assumed to be partially correlated in firm soil unless mentioned otherwise.

The random ground motion is assumed to be homogeneous stochastic process which is represented by Clough and Penzien double filter psdf given by two sets of filter coefficients representing the soft and firm soils respectively. For the soft soil , the coefficients are ω_g = 6.2832 rad/sec; ω_f = 0.62832 rad/sec ; $\xi_g = \xi_f = 0.4$, while those for the firm soils are ω_g = 15.708 rad/sec; ω_f = 1.5708 rad/sec; $\xi_g = \xi_f= 0.6$. The two psdfs corresponding to the two sets of filter coefficients are shown in Fig.5. The spatial correlation function used in the parametric study is given by Eqn.5 in which the value of c =2.0, Vs = 70 m/sec and Vs = 330 m/sec for the first and second psdfs respectively. The r.m.s (root mean square) ground acceleration is related to intensity of earthquake by empirical equation given by Eqn.7. Intensity of earthquake I_s in turn is related to magnitude of earthquake given by Eqn.9. The input for excitation is thus the intensity of earthquake for which $\sigma_{\ddot{u}_g}$ value can be calculated using Eqn.7. From $\sigma_{\ddot{u}_g}$ the value of S_0 defining the ordinates of the double filter psdf can be obtained using equation Eqn.1.

Parameter	Centre Span	Side Span
Deck length	L_2 = 335.28 m	$L_1 == L_3 = 137.16$ m
Deck Area	A_2 =0.32 m²	$A_1 = A_3$ =0.32 m²
Deck Depth	D_2 = 4.0m	$D_1 = D_3$ = 4.0
Modulus of Elasticity of deck	$E_2 = 2.0683 \times 10^{11}$ N/m²	$E_1 = E_3 = 2.0683 \times 10^{11}$ N/m²
Moment of Inertia of deck	I_2 = 1.131 m⁴	$I_1 = I_3 = 1.131$ m⁴
Tower Properties	L_t = 85.96m ; $E_t = 2.0683 \times 10^{11}$ N/m² ; A_t = 0.236 m²	
Cable Properties	Area of the cables (1 to 6)= 0.04, 0.016, 0.016, 0.016 and 0.04 m² Tension in cables (1 to 6)= 15.5 x 10⁶ , 5.9 x 10⁶, 5.9 x 10⁶, 4.3 x 10⁶, 5.9 x 10⁶ and 15.5 x10⁶ N/ m² Modulus of cables $E_c = 2.0683 \times 10^{11}$ N/m²	
Distributed mass of the bridge over half width deck		9.016 x 10³ Kg/m
Properties of flexible foundation	Radius of circular foundation = 3m Poission's ratio = 0.33 Density of the soil = 12Kn /m³	

Table 1. Structural data of the Harp Type Cable Stayed Bridge

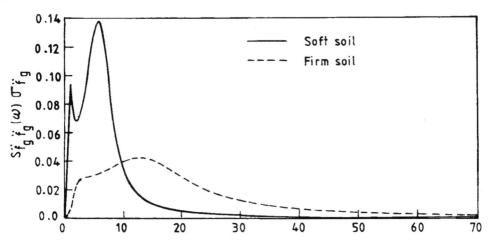

Fig. 5. Normalized PSDF of ground acceleration

Fig.6 shows the first five modes of the bridge corresponding to I_t/ I_d = 4.0. The first five frequencies and the corresponding nature of the mode shapes for the bridge is given in Table-2 for different I_t/ I_d ratios.

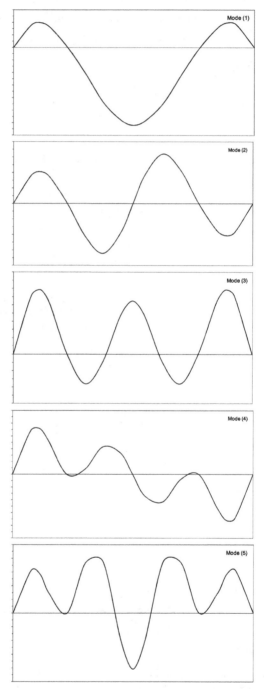

Fig. 6. First 5 mode shapes

Mode No.	Fundamental Frequencies (rad/sec)			Nature
	I_t / I_d =2.0	I_t / I_d =4.0	I_t / I_d =6.0	
1	2.838	2.983	3.072	Symmetric
2	3.247	3.551	3.716	Anti-symmetric
3	4.443	4.706	4.885	Symmetric
4	5.319	5.374	5.409	Anti-symmetric
5	6.429	6.450	6.467	Symmetric

Table 2. Fundamental frequencies of the bridge deck for different tower – deck inertia ratio.

6.1 Effect of the tower – Deck inertia ratio (I_t/I_d)

The first 5fundamental frequencies are obtained for different ratios of the tower – deck inertias i.e. 2, 4 and 6 respectively as shown in Tables-2. It is seen from the tables that with the increase of the tower – deck inertia ratio, the frequencies of the bridge deck increase. This is due to the fact that the increase in the tower stiffness increases the component of the vertical stiffness of the bridge provided by the cables.

6.2 Effect of the quasi-static component on the response

Total and relative displacement of the bridge decks obtained for the firm soil is shown in Tables-3. It is seen from the Tables that the contribution of the quasi-static component to the total response is significant for displacement where as it is small for bending moment.

Point No.	Relative		Total	
	Displacement (m)	Moment (t-m)	Displacement (m)	Moment (t-m)
1	0.0	0.0	0.0509	0.0
2	0.0961	1424.0	0.1108	1429.0
3	0.1040	1432.0	0.1184	1435.0
4	0.0831	727.0	0.0995	740.0
5	0.0	996.0	0.0509	1009
6	0.0906	715.0	0.1099	726
7	0.1621	974.0	0.1808	976.0
8	0.1886	702.0	0.2109	699
9	0.1957	1160.0	0.2164	1160.0

Table 3. Effect of the quasi – static part of the response on the r.m.s responses

6.3 Effect of barrier level on the reliability

Effect of the barrier level on the reliability of the bridge is shown in Figs. 7. The barrier level are taken as 15%, 20%, 25%, 30%, 33%, 40%, 50% and 70% of the yield stress assuming that the barrier level is the difference between yield stress and the pre-stress in the girder (deck). It is seen from the figures that the reliability increases as the barrier levels increases as it would be expected. However, the variation is not linear; it tends to follow an S shaped curve. For certain condition, the variation of reliability with barrier level may be very steep in the lower range of barrier levels. The same figures also compare between the reliabilities for firm and soft soils conditions. It is seen from the Fig.7 that the reliability for a particular barrier level is higher for firm soil. The difference between the two is considerably more at the lower end of the barrier level.

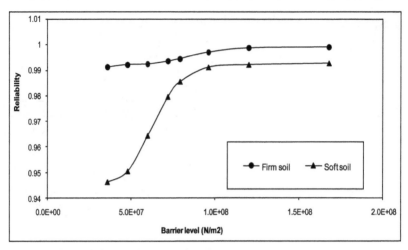

Fig. 7. Variation of Reliability with Barrier level

6.4 Effect of magnitude of earthquake on the reliability

The effect of the distribution of magnitude of earthquake on the reliability is shown in Fig.8. The figure shows the variation of reliability with barrier level for different distributions of the magnitude of earthquake obtained by exponential and gumbel distribution. It is seen from the figures that the reliability increases with the increase in beta values for exponential distribution. The difference between reliabilities obtained for two beta values considerably is more for lower values of barrier level. Above a certain value of beta, the reliability nearly approaches unity for all barrier levels. Further, Gumbel distribution provides much higher value of reliability as compared to Exponential distribution (for beta = 1.5). For soft soil condition (Fig.9), the effect of the distribution of the magnitude of earthquake is more produced.

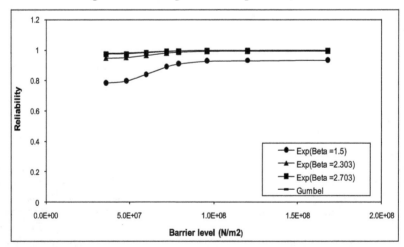

Fig. 8. Variation of Reliability with Barrier level for different distributions of magnitude of earthquake (Hard soil)

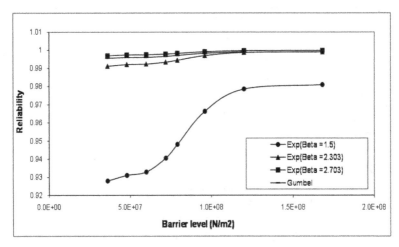

Fig. 9. Variation of Reliability with Barrier level for different distributions of magnitude of earthquake (Soft soil)

6.5 Effect of ratio between the three components of earthquake (Ru:Rv:Rw) on the reliability

The effect of this ratio on the variation of reliability with the barrier level is shown in Figs.10 and 11 for two angle of incidences of earthquake ($\alpha = 0^0$ and $\alpha = 45^0$). It is seen from the figures that the ratio has significant effect on this variation, especially at the lower end of the barrier level.

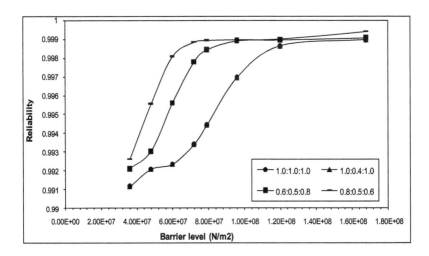

Fig. 10. Variation of Reliability for different ratios of Earthquake components (Alpha =0.0 degree)

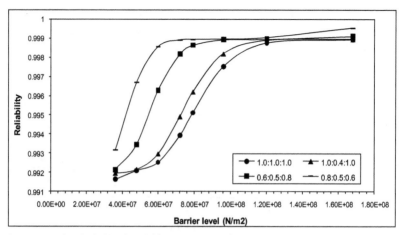

Fig. 11. Variation of Reliability for different ratios of Earthquake components (Alpha =45 degree)

6.6 Effect of angle of incidence on the reliability

Fig.12 show the effect of angle of incidence on the variation of reliability with barrier level. Three values of angle of incidence are considered namely, 0^0 i.e. major direction of earthquake is along the longitudinal axis of the bridge and the other two cases are having 30^0 and 70^0 angle of incidence with the longitudinal axis of the bridge. It is seen from the figures that 0^0 angle of incidence provides minimum reliability while 70^0 angle of incidence provides maximum reliability. This is expected because 0^0 angle of incidence produces maximum stress in the bridge.

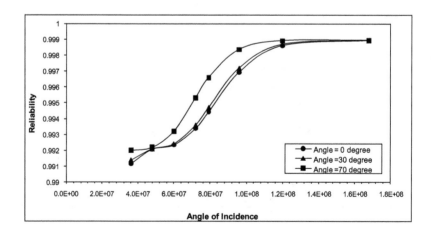

Fig. 12. Variation of Reliability for different Angles of Incidence

6.7 Effect of correlation of ground motion on the reliability

Fig.13 shows the variation of reliability with barrier level for three cases of ground motion that is fully correlated, partially correlated and uncorrelated. It is seen from the figures that fully correlated ground motion provides the highest reliability while uncorrelated ground motion gives the lowest value. The partially correlated ground motion gives reliability in between the two. This is the case because uncorrelated / partially correlated ground motion induces additional bending moment in the deck due to the phase lag of ground motion between different supports. The difference between the reliabilities for the three cases is nor very significant for the hard soil. However, the difference between them is significant for the soft soil (Fig.14). Further, the difference between the reliabilities are considerably reduced at the higher end of the barrier level.

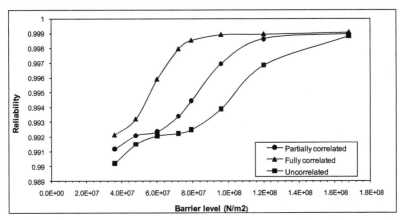

Fig. 13. Variation of Reliability with Barrier level for different degrees of correlation of ground motion (Hard soil)

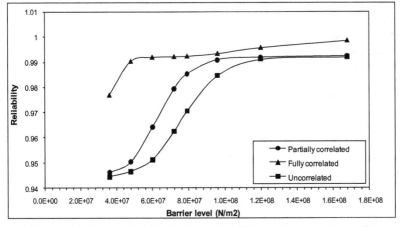

Fig. 14. Variation of Reliability with Barrier level for different degrees of correlation of ground motion (Soft soil)

6.8 Effect of duration of earthquake on the reliability

Fig.15 shows the variation of reliability with duration of earthquake for a barrier level of 33% . It is seen from the figures that reliability decreases mildly with the increase of duration of earthquake for both soft and firm soil. Thus, the duration of earthquake does not have significant influence on the reliability.

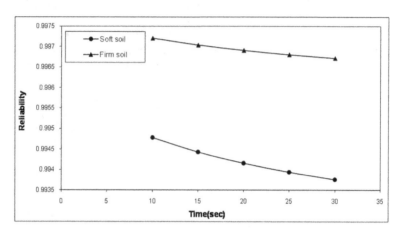

Fig. 15. Variation of Reliability with duration of Earthquakes (for a barrier level of 33%)

7. Conclusions

Reliability against first passage failure of cable stayed bridges under earthquake excitation is presented. The responses of cable stayed bridges are obtained for random ground motion which is modeled as stationary random process represented by double filter power spectral density function (psdf) and a correlation function. The responses are obtained by frequency domain spectral analysis. Conditional probability of first crossing the threshold level for a given RMS ground acceleration is obtained by using the moments of the psdf of the response. The RMS value of the ground acceleration is related to the magnitude of earthquake by an empirical equation, and the probability density function of the earthquake is integrated with the conditional probability of failure to find the probability of first passage failure. Using the above method of analysis, two cable stayed bridges are analyzed and probabilities of first passage failure are obtained for a number of parametric variations. The results of the numerical study lead to the following conclusions:

i.   The reliability against first passage failure increases sharply with the increase in barrier level in the lower range of its values.

ii. For the soft soil condition, the reliability is considerably less as compared to the firm soil condition.

iii. Gumbel distribution of the magnitude of the earthquake provides a very high estimate of reliability and gives values close to those obtained by exponential distribution with high values of the parameter β.

iv. Uncorrelated ground motion provides lower estimates of the reliability as compared to the fully correlated ground motion. The difference significantly more for the soft soil.

v.   The ratios between the components of ground motion have considerable influence on the reliability estimates. For soft soil condition, the difference between the reliability estimates, especially in the lower range of barrier level.

vi.   In the lower range of barrier level, considerable difference between reliabilities is observed for 0^0 and 70^0 angle of incidence with respect to the longitudinal axis of the bridge. For higher value of the barrier level, there is practically no difference between the two reliabilities.

vii. The duration of ground motion does not have significant influence on the reliability estimates.

viii. It is found that reliability decreases with the increase of average number of earthquake per year. The variation is nonlinear and more steep for the soft soil condition. For an average number of earthquake 0.5 per year , the reliability could be as 0.3.

8. References

[1] Abbas, A.M., and Manohar, C.S. (2007)."Reliability-based vector non-stationary random critical earthquake excitations for parametrically excited systems", Structural Safety, Vol.29(1), pp.32-48.

[2] Abbas, A.M., and Manohar, C.S. (2005)."Reliability-based critical earthquake load models. Part1:Linear Structures", Journal of Sound and Vibration, Vol.287, pp.865-882.

[3] Abbas, A.M., and Manohar, C.S. (2002)."Critical spatially-varying earthquake load models for extended structures", Journal of Structural Engineering, ASCE, Vol.29, pp.39-52.

[4] Ang, A.H.S., and Tang, W.H. (1975). " Probability concept in engineering planning and design basic principles", Vol. 1, John Wiley and Sons, New York.

[5] Abdel Ghaffar , A.M. and Nazmy, A.S. (1991). "3-D Nonlinear Seismic Behaviour of Cable Stayed Bridge" Journal of Structural Engineering, ASCE, Vol. 117, No. 11, pp. 3456-3476.

[6] Abdel Ghaffar, A.M., and Nazmy, A.S. (1987). "Effect of Three Dimensionality and Non-Linearity on the Dynamic and Seismic Behaviour of Cable Stayed Bridges", Bridges and Transmission Line Structures, Proceedings of Structures Congress, ASCE, Orlando, Florida, USA.

[7] Bolotin, V.V. (1965). "Statistical Methods in Structural Mechanics", STROIIDAT, Moscow (Translated by M.D. Friedman, Lockheed Missiles and Space Company).

[8] Crandall, S.H., and Mark, W.D. (1963). "Random Vibrations in Mechanical Systems", Academic Press, New York.

[9] Chern, C.H. (1976). "Reliability of Structure under Random Earthquake Motion", M.Eng. Thesis No. 952, Asian Institute of Technology, Bangkok.

[10] Clough, R.W., and Penzien, J. (1975). "Dynamics of Structures" , McGraw Hill, New York.

[11] Clough, R.W., and Penzien, J. (1993). "Dynamics of Structures" , McGraw Hill, New York.

[12] Harichandran, R.S., Hawwari, A. and Sweidan, B.N.(1996)."Response of Long-Span Bridges to Spatially Varying Ground motion", Journal of Structural Engineering, Vo.122(5), pp/476-484.

[13] Hong Hao (1998). "A Parametric Study of the Required Seating Length for Bridge Decks During Earthquake", Earthquake Engineering and Structural Dynamics, Vol. 27, pp. 91 –103.

[14] Hindy, A., and Novak, M. (1980). "Earthquake Response of Buried Insulated Pipes", Proc. ASCE, J. Engg. Mech. Div., Vol.106, pp.1135-1149.

[15] Kangan, Y.Y. (2002a)."Seismic moment distribution revisited: I. Statistical Result", Int. Journal of Geo-physical, Vol.148, pp.520-541.

[16] Kiureghian, A.D.(1996)."A coherency model for spatially varying ground motions", Earthquake Engineering & Structural Dynamics, Vol.29, pp.99-111.

[17] Konishi, I. (1969). "Safety and Reliability of Suspension Bridges", Proc. .First Int. Conf. on Structural Safety and Reliability (ICOSSAR), Ed. A.M. Freudenthal, Washington D.C.

[18] Lin, Y.K. (1967). "Probabilistic Methods in Structural Dynamics", McGraw-Hill, New York.

[19] Morris, N.F. (1974). "Dynamics Analysis of Cable –Stiffened Structures", Journal of the Structural Division, ASCE, Vol.100, No.ST5, pp.971-981.

[20] Mebarki, A., Lorrain, M., and Bertin, J. (1990). " Structural Reliability Analysis by a New Level – 2 Method: The Hypercone Method", Structural Safety, Vol. 9, No.1, pp. 91 – 103.

[21] Nigam, N.C. and Narayanan, S.(1994). "Applications of Random Vibrations",Narosa Publishing House, New Delhi.

[22] Penzien, J., and Watable, M. (1975). " Characteristics of 3-Dimensional Earthquake Ground Motions", Earthquake Engineering and Structural Dynamics, Vol. 3, pp. 365-373.

[23] Vanmarcke , E.H. (1975). "On the Distribution of the First –Passage Time for Normal Stationary Random Processes", Journal of the Applied Mechanics, Transactions of ASME, Vol.42, Paper No. 75-APMW-12, pp.215-220.

[24] Schueller, G.I. (1987) "Structures under Wind and Earthquakes Loading –Design and Safety Considerations" , A Short Course taught at Civil Engineering Department, Chulalongkorn University, Bangkok.

[25] Solomos ,G.P., and Spanos, P.T.D. (1982), "Structural Reliability under Evolutionary Seismic Excitation", Proc. First Int. Conf. on Soil Dynamics and Earthquake Engineering,U.K.(Published in Soil Dynamics and Earthquake Engineering „Vol.2, No.2).

[26] Soyluk, K., and Dumanoglu, A.A. (2000). "Comparison of Asynchronous and Stochastic Dynamic Response of a Cable Stayed Bridge", Engineering Structures, Vol. 22, pp. 435 – 445.

[27] Zhu, T. L. (1993). "A cell Technique for Computing the Failure Probability of Structural Systems", Computers and Structures, Vol. 46, No. 6, pp. 1001 – 1005.

4

Seismic Strengthening of Reinforced Concrete Buildings

Hasan Kaplan and Salih Yılmaz
Pamukkale University, Department of Civil Engineering
Turkey

1. Introduction

Many buildings have either collapsed or experienced different levels of damage during past earthquakes. Several investigations have been carried out on buildings that were damaged by earthquakes. Low-quality concrete, poor confinement of the end regions, weak column-strong beam behavior, short column behavior, inadequate splice lengths and improper hooks of the stirrups were some of the important structural deficiencies (Yakut et al., 2005). Most of those buildings were constructed before the introduction of modern building codes. They usually cannot provide the required ductility, lateral stiffness and strength, which are definitely lower than the limits imposed by the modern building codes (Kaplan et al., 2011). Due to low lateral stiffness and strength, vulnerable structures are subjected to large displacement demands, which cannot be met adequately as they have low ductility.

One of the first known examples of strengthening is the strengthening of Hagia Sophia by Sinan the Architect in 1573. With an insight casting light on the modern era, Sinan built buttress type shear walls around the mosque in order to reduce horizontal displacements of the building. When looked at the scientific studies that were carried out on this field, it is seen a research process started in 1950s (Whitney et al., 1955). In those years, infill wall tests which had started to be performed on single storey reinforced concrete frames continued with several structural types and strengthening methods.

Deficiencies that emerge in reinforced concrete buildings in terms of stiffness, strength, ductility and redundancy led to studies intended to strengthen buildings against earthquakes. The strengthening methods used today are intended to improve one or some of the behavioral characteristics of buildings listed above. Methods for the strengthening of buildings may basically be categorized into two main groups: System based strengthening and member based strengthening (Moehle, 2000). In the system based strengthening methods, a structural system is modified by adding members such as reinforced concrete shear walls, mainly improving the strength and stiffness characteristics of the system (Jirsa & Kreger, 1989; Albanesi et al., 2006). As to the member based strengthening methods, it is aimed at ensuring an improvement in the ductility of a system by means of enhancements made to those members with inadequate capacity or ductility. In these methods, it can be considered that there are no significant changes occurring in strength and stiffness characteristics of the load-bearing system.

Fig. 1. Hagia Sophia is strengthened with external buttress type walls by Sinan in 1573.

2. Design philosophy for strengthening

During earthquake motions, deformations take place across the elements of the load-bearing system as a result of the response of buildings to the ground motion. As a consequence of these deformations, internal forces develop across the elements of the load-bearing system and a displacement behavior appears across the building. The resultant displacement demand varies depending on the stiffness and mass of the building. In general, buildings with higher stiffness and lower mass have smaller horizontal displacements demands. On the contrary, displacement demands are to increase. On the other hand, each building has a specific displacement capacity. In other words, the amount of horizontal displacement that a building can afford without collapsing is limited. The purpose of strengthening methods is to ensure that the displacement demand of a building is to be kept below its displacement capacity (Yılmaz et al., 2011). This can mainly be achieved by reducing expected displacement demand of the structure during the strong motion or improving the displacement capacity of the structure.

In the system strengthening, new elements are added to a building to enhance its global stiffness. With an increase in the stiffness, the natural period of vibration of the building is to decrease. This, in turn, will result in a decrease in the amount of horizontal displacement that must be achieved by the building to resist earthquakes. When the building has enough stiffness, it will no longer able to achieve the amount of displacement which would cause it to collapse. Moreover, addition of new members to the building shall mostly increase the horizontal load capacity of the building as well (Canbay et al., 2003). The increased capacity will therefore require greater ground motions to allow the building to develop a yielding behavior. Thus, it can be said that the system strengthening does not only prevent collapsing but also delays structural damages. More clearly, the damage level of a building which is expected to sustain significant damages during a medium intensity earthquake may be reduced to minimum levels when it is strengthened by system strengthening techniques.

The element strengthening is a method based on the reinforcement of inadequate elements triggering the loss of stability due to the sustained damages without undertaking major changes in the load-bearing system of the building. In this method, negligible changes take place in the characteristics of a building such as stiffness and mass since no significant changes are made to the load-bearing system. Therefore, it must be considered that there will be no significant changes in the displacement demand after the member strengthening.

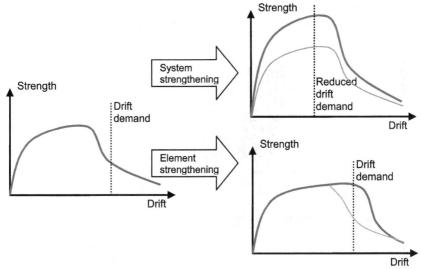

Fig. 2. Change in the displacement demand and capacity of the structure after strengthening

3. System based strengthening techniques

Most of the strengthening strategies have recently been based on global strengthening schemes as per which the structure is usually strengthened for limiting lateral displacements in order to compensate the low ductility (Sonuvar et al., 2004). In these methods causing a change in the global behavior of a building, as explained above, a behavioral change takes place when new members are added to the building. For a building which is currently used, it is important that the new members which are to be added to the structure are few in number and they are designed to ensure a significant increase in the load capacity and stiffness of the structure. It is well known by construction engineers that this target may most easily and economically be achieved through reinforced concrete and steel shear walls. Shear walls that are to be added to the building can be designed in the form of infill shear wall (Higashi et al., 1982) or external shear wall (Kaplan et al., 2009). Below are presented some of the most frequently applied system based strengthening methods.

3.1 Infill shear walls

Among the global strengthening methods, addition of RC infill is the most popular one. Many researchers have focused on the addition of infill RC walls and found that the installation of RC infill walls greatly improves lateral load capacity and stiffness of the structure. Even in cases of application to damaged buildings, the infill method can yield satisfactory results (Canbay et al., 2003; Sonuvar et al., 2004).

In the strengthening method with infill walls, the existing partition walls in the building are removed and high strength reinforced concrete shear walls are built instead (Figure 3). In such a strengthening application, the shear walls bear majority of the earthquake loads and limits the displacement behavior of the building while the frame system resists very low amounts of the earthquake loads. Reinforced concrete infill walls can also be used as partial walls and wing walls (Higashi et al., 1982; Bush et al., 1992). Door and window openings may also be provided in these walls to allow the building to deliver its architectural functions although they reduce stiffness and strength of the wall. In such applications preventing the construction of infill walls in the form of a full-fill wall, many experimental studies indicated that these walls develop more brittle damages compared to full-fill walls.

Fig. 3. Infill wall application in an RC building

As seen in Figure 4, columns of the existing building constitute the end regions of infill walls. Well confined end regions of reinforced concrete walls are considerably important in terms of the ductility of the wall. However, at the end regions of the strengthening walls, there generally exist inadequately confined columns with lower concrete strength. In that case, the end region must be confined in order to enhance ductility of the wall. Figure 4 presents various alternatives for confining end zones. Confined end region can be both made by jacketing of the existing column and can also be formed in the infill wall. If confinement and material characteristics of the existing columns are relatively good, the confined end region may be provided in the infill wall (TEC, 2007). In case the characteristics of end column are poor, it should be preferred to jacket the existing column.

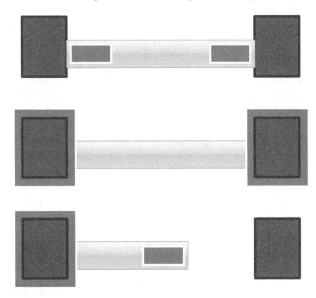

Fig. 4. Formation of the end zones in full or partial shear walls

Minimum reinforcement and anchor requirements for infill walls are given in Figure 5. It is also considerably important to ensure the continuity of longitudinal reinforcement in a manner that allows the wall to achieve an adequate moment capacity. Reinforcement continuity can be maintained by means of continuous anchors which are installed across beams and extended to both storeys. Anchors that are both installed across the foundation and the beams and fastened to the longitudinal reinforcements must have an enough lap splice length and embedment depth. If beam height is big, it would be difficult, or even impossible to drill the beam along its height and to install anchors. In such cases, wall can be constructed by breaking the beam or the longitudinal reinforcements can be placed on the sides of the beam by making the infill walls wider than the beam. In that case, anchored crossties must be installed on the lateral faces of the beam in order to restrict the buckling length of the reinforcement by fastening to the lateral face of the beam. In Figure 6, it is seen that with such anchors, there is a decrease in the buckling length of the longitudinal reinforcement extending along the sides of a beam. Anchored crossties can either be installed on both sides of a beam or be installed on each face separately.

Reinforced concrete walls to be added to the structure must be placed in a manner that it is not caused torsional effects on the structure and irregularities in the structure are eliminated, as observed in the design of new buildings. Some appropriate and inappropriate shear wall layouts are presented in Figure 7. The added walls must be regularly distributed across the plan. Unlike the design of existing building, a special example for strengthening is seen in Figure 7, at the bottommost. Although new walls are seemingly added in a symmetric and regular way to a building which have 2 irregular walls made from S220 grade steel, this layout is actually not appropriate. Since stiffness of the walls are close to each other, it is obvious that, in this building which is initially displaced without buckling, old walls will first reach the yield moment. There will be no enhancements in the new walls and the building will undergo twisting until the old walls lose their bearing capacity since there is no force balancing the earthquake loads that impact on the center of mass after these shear walls are failed. As it can be seen, in this seemingly correct design, the old walls are collapsed without any yielding in the new high capacity walls. However, the alternative of this seemingly appropriate incorrect design has a more regular behavior even though it exhibits a more irregular layout. Although some buckling is expected to occur across the elastic zone, the total capacities of those walls placed on the left- and right-hand sides of the structure are close to each other and will approximately yield at the same time. Any resultant accidental buckling is compensated through a shear wall placed in the middle zone.

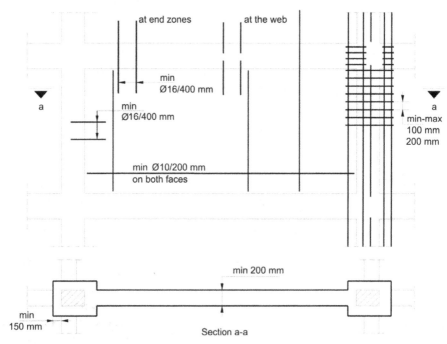

Fig. 5. Minimum reinforcement and anchor requirements for infill walls

3.2 External shear walls

Although the use of shear walls becomes widespread due to the fact that they are effective strengthening elements, they are also known to result in some difficulties hence they require a great deal of demolition and construction works in the existing structure. Application of external shear walls is an approach introduced to diminish such difficulties (Sucuoglu, 2006). In this approach, shear walls are applied to the external facade of a building without demolishing the existing infill walls. In that case, the shear wall can be placed in parallel with or perpendicular to the existing frame members.

Fig. 6. Wide infill walls and anchored crossties confining the longitudinal reinforcements

In case the shear walls are located perpendicular to the building façade (Kaltakci et al., 2008), large openings are needed in the building facade. The shear walls installed function like a buttress. In cases where pile foundation is not applied for the shear wall foundation, such shear walls are effective in only one direction. In order to create a positive effect on earthquake resistance of the building, they must be installed at opposite facades. This increases the required amount of shear walls and costs. For all these reasons, external shear walls that located perpendicular to the external facade are not preferred in the application.

The more preferred form of the external shear walls is the application where they are installed in parallel with the building façade (Yılmaz et al., 2010c, 2011). An example of the application of a shear wall located in parallel with the building facade is seen in Figure 8. Design and application of external shear walls are easier compared to that of infill walls. End zones of shear walls can be detailed in itself. These shear walls may also be manufactured as prefabricated panels and assembled in place (Yılmaz et al., 2010a, 2011; Kaplan et al., 2009).

Contrary to various advantages of external shear walls, there is a significant difference between external shear walls and infill shear walls. External shear walls shall not make a positive contribution to the frame strength when anchors connecting the external shear walls to the existing frame are damaged. However, even if infill shear walls are designed inadequately or applied incorrectly, it is apparent that they will have a bracing effect on the frame (Ohmura et al., 2006). In reinforced concrete design of external shear walls, it should be conformed to those guidelines that must be considered when designing reinforced concrete shear walls. What has importance is to correctly connect the external shear wall to the existing frame (Yılmaz, et al., 2010a, 2011). Although there is no methods for this provided in the applicable codes, from the experimental studies carried out by the authors, it was found that anchors can exhibit adequate performance when the design anchor shear force is determined in accordance with the capacity design principles.

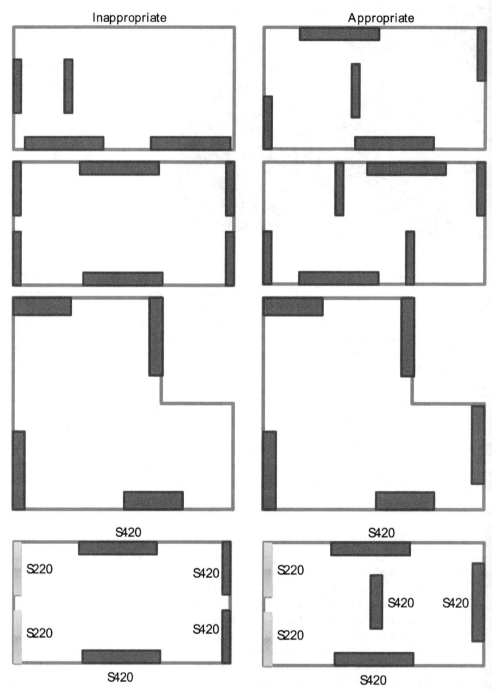

Fig. 7. Inappropriate and appropriate shear wall layouts

Fig. 8. An external shear wall strengthened building

The load transfer the external shear wall and the existing structure by the external shear wall anchors is schematically presented in Figure 9. To design anchors of the external shear wall, it is first determined the shear wall's ultimate moment capacity (M_p). The design shear force of the anchors applied to beams at the storey level, V_{ab} is calculated for i^{th} storey by the equation 1 given below. For calculations made for the last storey, it is taken that $V_{d,i+1}=0$. This force will give the anchor force required to transfer the necessary shear force to the shear wall at the relevant storey to allow the shear wall to reach its capacity at its bottom.

$$V_{ab,i} = (V_{d,i} - V_{d,i+1}) \cdot \frac{M_p}{M_d} \qquad (1)$$

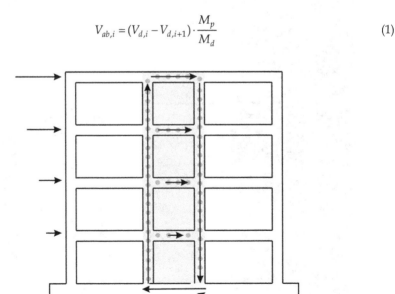

Fig. 9. Load transfer between the structure and external shear walls by anchors

After the storey anchor design obtained, contribution of the anchors to the shear wall moment capacity is established. When M_{ab} moment created by V_{ab} forces at the bottom of shear wall is subtracted from the ultimate bearing capacity of the shear wall, the residual moment (M_{ac}) obtained is expected to be carried by a force couple which is to be created by the anchors installed in the columns (Equation 3).

$$M_{ab} = \sum_{i=1}^{n} V_{ab,i} \cdot H_i \tag{2}$$

$$M_{ac} = M_p - M_{ab} \tag{3}$$

Experimental studies carried out indicated that an increase in the design anchor shear force for the last storey by 25 to 30 % has positive effects in terms of behavior (Yılmaz, et al., 2010a). In general, anchor damage which is to be created at the top storey subjected to the maximum anchor shear force results in a significant loss in the bearing capacity of the shear wall and causes the anchors at the lower storeys to be damaged or leads to serious damages at the top storey level (Figure 10).

Fig. 10. A damaged upper storey due to poor anchor design

3.3 Steel bracing

Steel bracing for RC frames has also been used to reduce drift demands. Bracing can either be implemented inside the frame (Masri & Goel, 1996) or applied from outside of the system (Bush et al., 1991) like RC walls. Post-tensioning can also be applied to bracing elements (Gilmore et al., 1996). In either case, steel bracing offers more suitable solutions in aesthetical terms for numerous applications. Although its application inside the building is not easy for those buildings with small openings, it particularly allows easy installation across the axes on external facades (Gorgulu et al., 2011). Research on various types of bracing styles is available in the literature (Perera et al., 2004; Ohmura et al., 2006). Architectural characteristics and functionality can be less disturbed by using an appropriate bracing style. Figure 11 presents the use of buttress type steel shear wall constructed on the building's external facade as a different example.

Design of steel elements must be made in conformity with the details specified in steel standards and codes. For the connection between the existing structure and steel members, the anchor design principles given in the subsequent section may be utilized. Another mode of damage that must especially be considered in the buttress type steel shear walls is to be the out-of-plane buckling of the compression elements of the buttress type shear wall placed. If possible, it is recommended to install lateral supports at storey levels to prevent lateral buckling. Use of such elements may also economize design of shear walls to a certain extent.

Fig. 11. Buttress type external steel shear wall (Photo by Dr. Yavuz Selim Tama)

3.4 Infill strengthening

Another method that can be used apart from adding new ductile shear walls to a building with an inadequate earthquake resistance is to improve both capacities and ductility of the existing brittle partition walls which are constructed between columns and beams in the form of fills. Those walls of which effects on the behavior are not considered during structural design can generally produce a certain bracing effect and make a positive contribution to the behavior. There are too many uncertainties regarding the behavior of these members and they are not likely to produce a desired bracing effect at all times. Partition walls may also lead to brittle damages across the surrounding columns by exhibiting many different behaviors. In addition, these walls are the members that are first damaged and lose their bearing capacity in a building under earthquake loads. A wide range of methods were developed to enhance capacity and ductility of these walls. Some of them include strengthening with mesh reinforcement, strengthening with precast panel and strengthening with FRP materials. The infill walls which are strengthened with those methods provide a bracing effect to the reinforced concrete frame for a longer period of time, restrict the building's displacement and produce a similar effect like the shear walls (Baran, 2005; Frosch et al., 1996). However, decrease in force is more brittle since the strengthened infill walls cannot display a bending behavior while the strengthening shear walls exhibits a bending behavior and reach their capacity in a ductile manner. Design engineer is expected to consider that these methods bring out a more brittle solution compared to those structures strengthened with reinforced concrete or steel shear walls.

4. Element-based strengthening

Element-based strengthening approach is the modification of deficient elements to increase ductility so that the deficient elements will reach their limit states in a ductile manner when subjected to design events. However, this strategy is more expensive and harder to implement in cases of many deficient elements which is the reason that the global strengthening methods have been more popular than element strengthening. Effective results can be obtained by using such methods in buildings with a limited number of deficient elements along with the global strengthening methods.

4.1 RC jackets
One of the most frequently used methods for strengthening of the reinforced concrete columns is the reinforced concrete jacketing (Figure 12). Jacketing which can be defined as the confinement of the column with new and higher quality reinforced concrete elements may be implemented for various purposes based on the type of deficiencies that the structural member has. Columns subjected to brittle damages can be jacketed in order to enhance resistance against shear and/or axial loads. In that case, although the purpose of jacketing is only to increase axial load or shear strength, some changes will also occur in the bending stiffness and moment capacity of the member after the jacketing application. By considering these changes during the jacketing design, the jacketed section is ensured to achieve adequate shear and axial load strength.

Except for such brittle damages, the jacketing is applied for elements with inadequate bending capacity or ductility. By this way, strength of the columns displaying a splice failure as a result of bending can also be improved. Jacketing of the columns is to produce the best result if it is implemented at 4 sides of the column. Where necessary, confinement at 3 sides can also provide adequate performance. However, it is not generally recommended to implement the jacketing at 1 or 2 sides. Because, with such jacketing applications, no significant changes take place in the confinement characteristics of the member.

Fig. 12. An RC jacketing application (Photo by Ahmet Sarışın)

Although the reinforced concrete jacketing can technically be applied for all the structural members, a jacketing application which will increase the beam's bending capacity is not advised since it may cause the strong beam-weak column formation resulted by an increase in the capacity of beam members.

4.2 Steel jackets

Jacketing with steel elements is a practical method used frequently for various applications. A typical steel jacketing application is presented in Figure 13. Steel jacketing can readily be used to especially enhance the shear strength of reinforced concrete elements. Located at the corners of an element, L-profiles are coupled by means of steel plates and confined. With the maintenance of continuity between storeys, steel jacketing can also be used to increase the bending strength. Also, the maintenance of adequate strength between the steel element and reinforced concrete element is inevitable for the improvement of bending capacity.

4.3 Fiber reinforced polymers

In recent years, use of Fiber Reinforced Polymer has considerably become widespread in strengthening applications. Fiber polymer fabrics that can be used to improve bending, shear and axial capacities of the columns and beams may be manufactured from various materials such as carbon, glass and aramidWithout an increase in the volume of the strengthened member, significant improvements can be achieved in the capacity and ductility characteristics of the element. In Figure 14, beam strenghthening for a parking structure is shown. Due to space restrictions in this structure, it was not feasible to utilize classical element strengthening techniques. These materials may practically be used for numerous purposes such as enhancement of the flexural capacity of floor slabs and improvement of shear capacity of beams, columns or shear walls.

Fig. 13. Steel jacketing applied to RC columns (Photo by Dr. Ali Haydar Kayhan)

In case FRP material is used like a longitudinal reinforcement, the additional flexural capacity produced can easily be found by a simple calculation of cross-section. Calculations for determining its contribution to the shear capacity is not very different from the conventional reinforced concrete calculations.   To measure enhancement in the axial capacity and change in the ductility of the member, more complex calculations beyond the limits of basic reinforced concrete knowledge are needed.

Fig. 14. FRP confinement of a beam (reprinted with the permission of FYFE Europe)

5. Connections between new and existing elements: Anchors

5.1 Design

Anchor design should primarily be made by considering different modes of behavior that the anchor members can exhibit. The most general approach to determine the anchor strength is the capacity approach given in ACI318 Appendix D (2005). Although the approach suggested in ACI318 is recommended for cast-in-place and mechanical anchors, it was found from the experimental studies that it can also be used for most of the chemical anchors. Both in shear or tensile loads, the ultimate capacity could be calculated for the different collapse modes in the ACI318 method and the ultimate capacity would be determined taking the collapse mode with the lowest strength into consideration.

Possible failure modes that the anchor elements are likely to sustain under tensile forces are shown in Figure 15 (Yılmaz & Özen, 2010a). In the ACI method, capacity values for the failure modes indicated below are calculated separately or determined empirically. After the relevant capacity values are established for the steel failure (N_{sa}), debonding failure (N_p), cone failure (N_{cb}) conditions, the tensile capacity of the anchor element is considered to be the lowest one of these values. While possibility of the occurrence of a cone failure in shallow anchors which are implemented for those elements with lower concrete strengths increases, increased diameter of the steel also raises that possibility. Debonding failure is affected by concrete strength and diameter of the steel as well as numerous unknown factors related to the application and is definitely recommended to be determined experimentally.

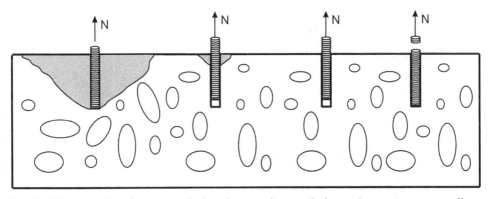

Fig. 15. Failure modes of post-installed anchors under tensile forces (concrete cone, small cone with debonding, debonding, steel failures)

Similar to tensile failure modes, in ACI318 Appendix D, three different failure modes were defined upon reaching the anchor shear capacity: steel failure (V_{sa}), concrete pryout failure (V_{cp}) and concrete breakout failure (V_{cb}) of near edge anchors. These modes of collapse are shown in Figure 16. In situations where the anchor bars is closer to the edge, the ultimate concrete cone capacity generally governs the anchor capacity and where it is further away from the edge, the shear capacity of the anchor bar is the main determining factor. For shallow or high diameter anchors, concrete pryout failure may also govern the behavior (Çalışkan et al., 2010).

Tensile or shear capacity determined with the abovementioned approach should be reduced by a strength reduction coefficient for design purposes. Various coefficients are listed in ACI318 depending on the environmental conditions and workmanship quality. These reduction coefficients vary between 0.45 and 0.75 depending on the loading type, failure mode and reliability of the application. Details of the formulations are to be studied from related codes or standards.

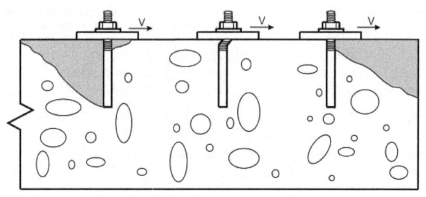

Fig. 16. Failure modes of post-installed anchors under shear forces

In some studies carried out recently on low strength concretes, it is suggested to use a reduction coefficient depending on the diameter of reinforcement when the anchor capacity is governed by steel failure mode per ACI318 formulations. Accordingly, the strength of reinforcements with larger diameters (bar diameters greater than 16mm) should be reduced, as seen in Figure 17. Research on this area is not completed yet. Designers are recommended to use high diameter bars with concern.

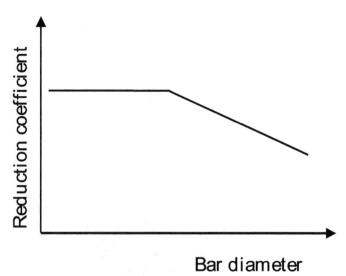

Fig. 17. Strength reduction for steel failure mode depending on bar diameter

An important question for the design engineer is which failure modes will be allowed or not allowed. That is to say, it must become apparent that anchor capacity would allow failure modes such as cone or pryout which are of brittle failure modes or always a ductile failure arising from anchor bar should take place. The answer to this question varies depending on the usage mode of the anchor. For example, let us consider anchors which are overlapped to the longitudinal reinforcements at the bottom of shear wall. Since the shear wall shall achieve bending capacity due to tensile forces developed in these anchors, the shear wall cannot have a ductile bending behavior as expected if collapse takes place in a brittle failure mode in these anchors. For ductile behavior of the shear wall, these anchors must exhibit a yield behavior without formation of debonding or concrete cone. In case that an anchor with a diameter larger than the longitudinal reinforcement of shear wall is installed, the anchor capacity can be determined by brittle failure modes, but in that case, the anchor strength must not be lower than that of the longitudinal reinforcement connected so that the longitudinal reinforcement can display a ductile tensile yielding. When the shear wall reaches the bending capacity, capacity of other anchors that are found not to develop a yield mechanism can also be allowed to be designed according to brittle failure modes. In that case, brittle anchor failures will never occur since the anchor cannot reach its capacity.

5.2 Application

Performance of the elements connecting the old and new members mostly has an effect on the performance of the strengthened system. Thus, connection elements used in the strengthening must carefully be designed and implemented. Since majority of the application is comprised of anchor elements, this section is focused on some topics related to such elements. The most frequently used anchor type in strengthening applications is chemical anchors. Being readily applied in field, these anchors are also substantially cost-effective.

However, many reinforced concrete elements to which anchors are applied for the strengthening works have a very poor concrete quality and some of these elements have been seriously damaged during the service life of the structure. However, it is crucial to consider failure conditions of these elements during anchor installation. It is also important that holes to be used for anchor installation must be bored so as to cause the lowest level of damage to these elements. Figure 18 presents some incorrect applications made on low strength concretes.

One of the most important defects that can occur during the installation of chemical anchors is to carry out installation by applying chemical material around the steel bar only without adequately filling the holes with adhesive. Tensile tested specimen of an anchor bar installed in this way is presented in Figure 19. It is seen that epoxy on the surface of reinforcement is not in contact with a vast section of the hole. It is obvious that such an anchor will prevent the strengthening element from reaching its capacity.

Hole must be bored in a manner that the hole diameter is at least 4~5 mm higher than the bar diameter and between 120% and 150% of the bar diameter. Before the hole boring process, a reinforcement detection device must be used, otherwise lots of trials may be needed to bore holes and this results in unnecessary damages (Figure 20). In addition, where applicable, anchor bars must not be installed in defective zones. It should be considered that anchors that are near the edge of reinforced concrete element will have a decreased efficiency. Some anchors implemented across inappropriate locations are shown in Figure 20.

Fig. 18. Poorly implemented anchor examples (A, B: anchors implemented on defective surfaces; C: Anchor implemented near a crack; D: Anchor fracturing the concrete during the tensile test)

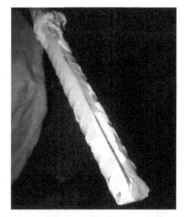

Fig. 19. Interlocking surface in an anchor installed into a hole which is not filled with epoxy (Photo by Dr. Halil Nohutcu)

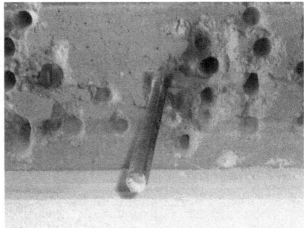

Fig. 20. Inappropriate anchor layouts

6. Conclusion

A brief summary of the available literature on seismic strengthening methods is presented in this chapter. In general, the structural engineer has alternatives of element based and system based methods. Element based techniques are more economical solutions when local problems are the main reason of the strengthening decision. If the problems related with the global ductility, stiffness or strength of the structure are the main concern, global strengthening techniques are more advantageous. Both approaches can also be utilized if needed.

On the other hand, performance of the elements connecting the old and new members mostly has a vital effect on the performance of the strengthened system. Thus, connection elements used in the strengthening must carefully be designed and implemented. Since majority of the application is comprised of anchor elements, this section is focused on some topics related to anchors.

7. Acknowledgment

Authors greatly acknowledges Dr. Halil Nohutcu, Dr. Nihat Cetinkaya, Dr. Özlem Çalışkan, Ahmet Sarisin, and Prof. Dr. Ergin Atimtay for their efforts and contributions in experimental program on external shear walls. Dr. Yavuz Selim Tama, Dr. Ali Haydar Kayhan are also acknowledged for providing some photos on strengthening. Authors supervises several strengthening projects and carried out many research projects. Funding institutions (TUBITAK, DPT and Pamukkale University) and organizations (Denizli, Mugla and Kütahya Governorships) are also acknowledged for their valuable support.

8. References

ACI Committee 318. Building code requirements for structural concrete and commentary (ACI 318M-05). American Concrete Institute, 2005.

Albanesi T., Biondi S., Candigliota E. and Nuti C. Experimental analysis on a regular full scale infilled frame. Proceedings of the First European Conference on Earthquake Engineering and Seismology, Geneva, 2006, Paper No. 1608.

Baran M. Precast Concrete Panel Infill Walls for Seismic Strengthening of Reinforced Concrete Framed Strcutures. PhD thesis, Middle East Technical University, Ankara, 2005.

Bush T. D., Wyllie L. A. and Jirsa, J. O. Observations on two seismic strengthening schemes for concrete frames. Earthquake Spectra, 1991, 7, No.4, 511-527.

Çalışkan, Ö., Yilmaz, S., Kaplan, H., "Shear Capacity of Post-Installed Anchors According to ACI318 and TS500" 9th International Congress on Advances in Civil Engineering, 2010, Trabzon, Turkey.

Canbay E., Ersoy U. and Ozcebe G. Contribution of reinforced concrete infills to seismic behavior of structural systems. ACI Structural Journal, 2003, 100, No.5, 637-643.

Frosch R. J., Wanzhi L., Jirsa J. O. and Kreger M. E. Retrofit of non-ductile moment-resisting frames using precast infill wall panels. Earthquake Spectra, 1996, 12, No.4, 741-760.

Gilmore A. T., Bertero V. V. and Youssef N. F. G. Seismic rehabilitation of infilled non-ductile frame buildings using post-tensioned steel braces. Earthquake Spectra, 1996, 12, No.4, 863-882.

Gorgulu, T., Tama, Y.S., Yilmaz, S., Kaplan, H., Ay, Z., Strengthening of Reinforced Concrete Structures With External Steel Shear Walls, Journal of Constructional Steel Research, 2011, doi:10.1016/j.jcsr.2011.08.010.

Higashi Y., Endo T. and Shimizu Y. Effects on behaviors of reinforced concrete frames by adding shear walls. Proceedings of the Third Seminar on Repair and Retrofit of Structures, Michigan, 1982, pp. 265-290.

Jirsa J. and Kreger M. Recent research on repair and strengthening of reinforced concrete structures. Proceedings of the ASCE Structures Congress, California, 1989, 1, 679-688.

Kaltakci, M.Y., Arslan, M. H., Yilmaz, U. S. and Arslan, H. D. A new approach on the strengthening of primary school buildings in Turkey: An application of external shear wall, Building and Environment, 2008, 43, No.6, 983-990.

Kaplan, H., Yılmaz, S., Cetinkaya, N., Atımtay, E., Seismic strengthening of RC structures with exterior shear walls, Sadhana - Academy Proceedings in Engineering Science, 2011, 36(1), 17-34.

Kaplan, H., Yılmaz, S., Cetinkaya, N., Nohutcu, H. and Atımtay, E. Gönen, H. A New Method for Strengthening of Precast Industrial Structures. Journal of The Faculty of Engineering and Architecture of Gazi University, 2009, 24 No.4, 659-665, (in Turkish).

Masri A. and Goel S. Seismic design and testing of an RC slab-column frame strengthened with steel bracing. Earthquake Spectra, 1996, 12, No.4, 645-666.

TEC-2007, Ministry of Public Works and Settlement, Turkish Earthquake Code-2007: Specifications for buildings to be built in seismic areas. Ankara, Turkey (in Turkish), 2007.

Moehle J. P. State of research on seismic retrofit of concrete building structures in the US. Proceeding of US-Japan Symposium and Workshop on Seismic Retrofit of Concrete Structures - State of Research and Practice, USA, 2000.

Ohmura, T., Hayashi, S., Kanata, K. and Fujimura, T., Seismic retrofit of reinforced concrete frames by steel braces using no anchors. Proceedings of the 8th National Conference on Earthquake Engineering. California, 2006.

Perera, R., Gómez, S. and Alarcón, E. Experimental and analytical study of masonry infill reinforced concrete frames retrofitted with steel braces. ASCE Journal of Structural Engineering, 2004, 130, No.12, 2032-2039.

Sonuvar M. O., Ozcebe G. and Ersoy, U. Rehabilitation of reinforced concrete frames with reinforced concrete infills. ACI Structural Journal, 2004, 101, No.4, 494-500.

Sucuoglu H., Jury R., Ozmen A., Hopkins D. and Ozcebe G. Developing retrofit solutions for the residential building stocks in Istanbul. Proceedings of 100th Anniversary Earthquake Conference, California, 2006

Whitney, C.S., Anderson, B.G. ve Cohen, E., Design of Blast Resistant Construction for Atomic Explosions, ACI Structural Journal, 1955, 26(7):589-683.

Yakut A., Gülkan P., Bakır B.S., Yılmaz M.T., Re-examination of damage distribution in Adapazarı structural considerations. Engineering Structures, 2005, 27, No.7, 990-1001.

Yılmaz, S., Cetinkaya, N., Nohutcu, H., Caliskan, O., Çırak, İ.F., Experimental Investigation on Anchor Applications for External Shear Walls, Technical Report, 192 p, 2010a, Denizli.

Yılmaz, S., Kaplan, H., Çalışkan, Ö., Kıraç, N., "Cyclic Shear Resistance of Epoxy Anchors Bonded to Low Strength Concrete" 14th European Conference on Earthquake Engineering, Paper no. 510, 2010b, Ohrid, Macedonia.

Yılmaz, S., Özen M.A., "Tensile Strength of Chemical Anchors Embedded to Low and Normal Strength Concrete" Pamukkale University, Report no: 2009FBE025, 131 p. 2010a, Denizli (in Turkish).

Yılmaz, S., Özen, M.A., "Strength of Chemical Anchors Embedded to Low Strength Concrete" 14th European Conference on Earthquake Engineering, Paper no. 513, 2010b, Ohrid, Macedonia.

Yılmaz, S., Tama, Y.S., Kaplan, H., "Design and Construction of External Precast Shear Walls for Seismic Retrofit" 14th European Conference on Earthquake Engineering, Paper no. 1238, 2010c, Ohrid, Macedonia.

Yılmaz, S., Kaplan, H., Tama, Y.S., Çalışkan, Ö., Solak, A., "Experimental program on design and application of external retrofit walls for low ductility RC frames" 4th International Conference on Advances in Experimental Structural Engineering, 2011, Ispra, Italy.

5

Assessment and Rehabilitation of Jacket Platforms

Mohammad Reza Tabeshpour[1], Younes Komachi[3] and Ali Akbar Golafshani[2]
[1]Mechanical Engineering Department,
[2]Civil Engineering Department
Sharif University of Technology, Tehran
[3]Department of Civil Eng., Pardis Branch, Islamic Azad University, Pardis
Iran

1. Introduction

The most important recent earthquake showed the importance of seismic assessment of both onshore and offshore structures (Takewaki et al., 2011). Strong earthquake can cause damages to engineering structures. Many strong earthquakes normally take place in offshore such as 2011 Japan earthquake (Moustafa, 2011 and Takewaki et al., 2011) can cause severe damage to offshore structures. The steel jacket structure is a kind of fixed offshore platform that is suitable for construction in water depth from a few meters to more than 100 m. Compared to regular structures, a jacket offshore platform is a complicated system and is composed of many parts include structural and nonstructural elements. Structural modeling includes two major division; structure and pile foundation. Fig. 1 shows parts of jacket offshore platform system and some of related researches.

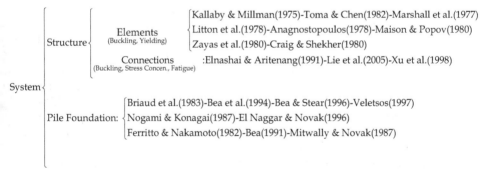

Fig. 1. Parts of jacket offshore platform system.

The major structural components of such an offshore platform are jacket, piles, and deck. A jacket structure which serves as bracing for the piles against lateral loads is fixed by piles driven through the inside of the legs of the jacket structure and into soil many tens of meters deep. The deck structure is fixed upon the jacket structure. Oceans in which offshore

platforms are built present a set of complicated and harsh environmental conditions. Dynamic loads including wind, wave, current, and earthquakes dominate the design of offshore structures. Venkataramana (1998) presented a time domain analysis of the dynamic response of a simplified offshore structure to simultaneous loadings by random sea waves and random earthquake ground motions. Kawano and Venkataramana (1999) investigate the dynamic response and reliability analysis of offshore structure under the action of sea waves, currents and earthquakes. The dynamic loads affect not only the routine operation of an offshore platform such as drilling and production activities, but also the safety and serviceability of the structure. Approximately 100 template-type offshore platforms have been installed in seismically active regions of the world's oceans. New regions with the potential for significant seismic activity are now beginning to be developed. Older platforms in seismic regions may have three areas of deficiency:
1.   Inadequate ground motions for original design.
2.   Structural framing which is not arranged or detailed for ductile behavior.
3.   Reduced capacity resulting from damage, corrosion or fatigue.
Many of these platforms are now beyond their original design life (20-25 years), as well. From the economic point of view the continued use of an existing installation will in many cases be preferable compared to a new installation. The assessment of existing platforms under environmental (wave, wind, current etc.) loads and probable future loads (earthquake) (Moustafa, 2011) is a relatively new process and has not yet been standardized as design is. This lack of standardization creates some difficulty in establishing performance requirements which must be developed depending upon the risks (i.e., hazards, exposures and consequences) associated with the future operations of the platform. The present criteria of the offshore structure standards for seismic assessment can be improved using building pre-standards.

Assessment of jacket platforms has rarely been studied. Krieger et al. (1994) describe the process of assessment of existing platforms. Petrauskas et al. (1994) illustrate assessment of structural members and foundation of jacket platforms against metocean loads. Craig et al. explain assessment criteria for various loading conditions. Ersdal (2005) evaluates the possible life extension of offshore installations and procedures of standards in this matter, with a focus on ultimate limit state analysis and fatigue analysis. Gebara et al. (2000) assess the performance of the jacket platform under subsidence and perform ultimate strength and reliability analyses for four levels of sea floor subsidence. The assessment process of building prestandards was studied also. Bardakis and Dritsos (2007) compared the criteria of FEMA-356 and GRECO (Based on the EC-8). Hueste and Bai (2007) described the assessment and rehabilitation of an existing concrete building based on the FEMA-356 procedure and criteria. Golafshani et al. (2009) suggested this idea that API procedures for seismic assessment of jacket platforms can be evaluated and improved with respect to building documents for the first time at 2006.

Fig. 2 shows some standards for offshore structures that include detailed procedures for the assessment of existing structures. Petroleum and natural gas industries Offshore Structures Part 1: General Requirements,' ISO (2002) is one of the most general accepted standards. A detailed assessment procedure for existing structures is found in ISO 19902. The Norwegian regulations (PSA 2004) refer to ISO 19900 (ISO 2002) for the assessment of existing structures. Other standards, like API RP2A-WSD (API 2000) and ISO/DIS 13822 (ISO 2000), also include detailed procedures for the assessment of existing structures.

API RP2A is one of the most useful standards for the design and assessment of offshore structures. Section 17 of this standard has recommendations for the assessment of offshore structures. The assessment criteria of this standard are based on the objective of collapse prevention of the structure under extreme earthquake conditions. The results of an assessment with API give information about the total structure's condition. This standard describes the rehabilitation objective globally and does not present a routine methodology for rehabilitation.

In the last decade, several building documents such as FEMA- 356 and ATC-40 were developed for the assessment and rehabilitation of these structures. In the FEMA-356 document seismic deficiencies identified using an evaluation methodology considering building performance at a certain seismic hazard. The FEMA-356 document developed an extensive assessment and rehabilitation procedure. This document not only has numerical criteria for assessment but also presents design procedures for rehabilitation.

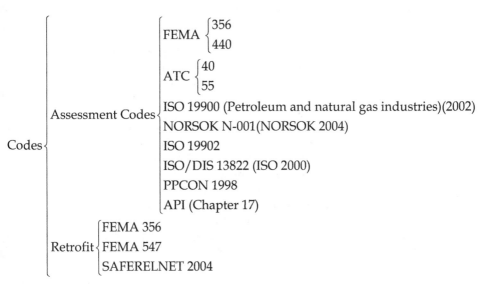

Fig. 2. Some standards for assessment and rehabilitation of jacket offshore platforms

The API approach is based on simple collapse mechanism investigation without describing the methodology in detail. However FEMA-356 consists of detailed processes for both seismic assessment and rehabilitation of building structures. However there are advices in API documents (API RP 2A) in order to seismic assessment of jacket platforms, but because of brief existing comments in this field, it is necessary to use more appropriate pre-standards for seismic assessment of these structures. For example Golafshani et al., (2009) compares FEMA-356 and API approach for assessment of jacket offshore platform structures. Komachi et al., (2009a) presented the performance based assessment of jacket platforms for seismic vulnerability.

Current methods for seismic upgrading of existing structures can be classified into two major groups: traditional and modern. Traditional methods aim to increase the strength and/or ductility of the structure by repairing/upgrading members. Nowadays there are some new technologies (for example seismic isolation and energy dissipation) for seismic protection of the structures. The passive control approach is of current concern to many researchers and there are several attempts exploring its application to offshore structures. Recently, although, there have been several studies for the effectiveness of active and passive control mechanisms in controlling the response of offshore platforms under wave loading.

Incorporation of energy dissipation systems in a traditional earthquake-resistant structure has been recognized as an effective strategy for seismic protection of structures (Soong and Dargush, 1997). New vibration control technologies have been applied to offshore structures in the following cases. Vandiver and Mitome (1979) used storage tanks as Tuned Liquid Damper (TLD) on a fixed platform to mitigate the vibration of the structure subjected to random wave forces. Kawano and Venkataraman (1992) and Kawano (1993) studied the application of an active tuned mass damper to reduce the response of platforms due to wave loading. Abdel-Rohman (1996) studied the dynamic response of a steel jacket platform with certain active and passive control due to wave-induced loading. Lee (1997) used stochastic analysis and demonstrated the efficiency of mechanical dampers for an offshore platform. Suneja and Datta (1998) demonstrated the efficiency of an active control system for articulated leg platforms under wave loading. Vincenzo and Roger (1999) developed an Active Mass Damper for suppression of vortex- induced vibrations of offshore structures. Chen et al. (1999) studied the response of a jacket platform installed with TLD due to earthquake loading. Ou et al. (1999) studied the response reduction of jacket platforms with a viscoelastic damper with respect to ice loads. Terro et al. (1999) developed a multi-loop feedback-control design as applied to an offshore steel jacket platform. Suhardjo and Kareem (2001) used both passive and active control systems for the control of offshore platforms. Ding (2001) studied the response reduction of jacket platforms with a viscous damper due to ice loads. Qu et al. (2001) presented a rational analytical method for determining the dynamic response of large truss towers equipped with friction dampers under wind-excitation and investigated the efficiency of friction dampers. Wang (2002) used Magnetorheological dampers for vibration control of offshore platforms for wave-excited response. Mahadik and Jangid (2003) studied the response of offshore jacket platforms with an active tuned mass damper under wave loading. Patil and Janjid (2005) studied the behavior of a platform with viscoelastic, viscous and friction damper for wave loads. Lee et al. (2006) studied the effectiveness of a Tuned Liquid Column Damper (TLCD), which dissipates energy by water flow between two water columns, for offshore structures and also, Ou et al. (2006) studied the application of damping isolation systems for response mitigation of offshore platform structures. Jin et al. (2007) studied the effect of Tuned Liquid Dampers (TLD) and found that the larger the ratio of water-mass to platform-mass, the higher the reduction of responses. Komachi et al., (2009) presented Friction Damper Devices (FDD) as a control system to rehabilitation existing jacket offshore platforms. Golafshani and Gholizad (2009) studied the performance of friction dampers for mitigating of wave-induced vibrations and used mathematical formulation to evaluate the response of the model. Yoe et al. (2009) used Tuned Mass Damper (TMD) for mitigation of dynamic ice loads.

The service life of an offshore structure can be doubled if the dynamic stress amplitude reduces by 15%. Few studies have reported on the effectiveness of the passive control systems using dampers in controlling the response of offshore platforms under a parametric variation studying the influence of important system parameters and comparative performance of dampers. In order to reduce possible damage to jacket offshore platforms in harsh marine environments, the necessity of carrying out further studies on developing efficient and practical vibration control strategies for the suppression of dynamic responses of existing offshore structures should be emphasized. In this chapter rough and global comments of API are compared with detailed method of FEMA. As an example seismic assessment of the existing 4 legged Service platform placed in the Persian Gulf is presented. A very useful method for rehabilitation of existing jacket platforms is damper. In this study, a Friction Damper Device (FDD) proposed by Mualla (2002) is used to mitigate the vibration of a typical fixed jacket offshore platform in Persian Gulf. The contents of this study mainly include the investigation of the influence of the damping system parameters on vibration control of offshore platforms under the actions of earthquake excitations. This chapter shows that FDD improves the structural behavior and performance of jacket platforms.

2. Assessment

2.1 API recommendations
The API is currently developing recommendations for the assessment of existing platforms including requirements for platforms subjected to hurricanes, storms, earthquakes and ice loading. These recommendations will likely focus on a demonstration of adequate ductility for platforms located in earthquake dominated regions. The focus towards ductility, or demonstrated survivability, under extreme earthquake conditions is based on the objective of prevention of loss of life and pollution. The performance criterion for assessment is essentially identical to that of the Design Level Earthquake (DLE) requirement for new designs.

The structures need to meet one of two sets of global structural performance criteria, depending on the platform's exposure category. In addition, local structural performance requirements for topside equipment and appurtenances must be met, independent of the platform's exposure category classification. In the case of high exposure platforms, they must be shown by rational analysis (Pushover or Nonlinear Time history) to remain globally stable under median ground motions representative of an earthquake with an associated return period of 1000 years. For lower exposure platforms located in areas with high seismic activity, a return period of 500 years must be selected.

2.2 FEMA recommendations
Performance based design (PBD) has been fully described in the guidelines published by FEMA and ATC. These documents do not have the force of codes but provide details of best practice for the evaluation and strengthening of existing buildings. These are continuing to be expanded as PBD becomes more widespread. In these standards a criterion such as drift is applied indirectly when the elements are assessed.

Four levels of building performance consist of Operational (O), Immediate Occupancy (IO), Life Safety (LS), and Collapse Prevention (CP) in increasing levels of damage is considered in FEMA. The Immediate Occupancy (IO) performance level requires that the building remain essentially functional during and immediately after the earthquake. The last performance level, Collapse Prevention (CP) will result in a building on the point of collapse and probably economically irreparable. The Life Safety level (LS) is the level usually implicit in codes and may also result in a building which is not economic to repair. The rehabilitation objectives are formed of combinations of earthquake hazard and building performance (consisting of structural and nonstructural performance levels).

FEMA sets a desirable goal for rehabilitation, a Basic Safety Objective (BSO) which comprises two targets:

1. Life Safety building performance at Basic Safety Earthquake 1 (BSE-1), the 475 year earthquake.
2. Collapse Prevention performance at BSE-2, the 2500 year earthquake.

Depending on the function of the structure, other objectives may be set. For example, an enhanced objective may set higher building performance levels at BSE-1 and BSE-2 for critical facilities such as hospitals. Component actions are classified as deformation or force-controlled. Table 1 shows some of the action types. A component acceptance criterion for force-controlled actions is based on the force and is independent of the performance level and components shall have lower-bound strengths not less than the maximum design forces. For deformation-controlled actions the criterion is based on the target performance level and in these components shall have expected deformation capacities not less than maximum deformation demands calculated at the target displacement.

Action		Force-Control	Deformation-Control
Braces in Tension and Compression			✓
Columns – Compression		✓	
Columns – flexure	$P < 0.5P_{cr}$		✓
	$P > 0.5P_{cr}$	✓	

Table 1. Type of action for some elements

FEMA provides modeling parameters and numerical acceptance criteria for beams, columns, and braces as a function of parameters such as diameter to thickness ratio. The leg and brace in the jacket act like a column and brace in the steel braced frame, respectively. Therefore, for assessment of these structures using FEMA, criteria of chapter five of this document will be used.

2.3 Comparison of API RP 2A and FEMA-356

In the Table 2 procedures of FEMA and API for assessment are compared to each other. In contrast to API that assesses structures globally, FEMA evaluates each member of the structure for assessment. This matter has some advantages in several manners:

	API RP 2A	FEMA-356
Structure	Jacket Platform	Building
Criteria for	Total structure	Members and structure
Return Period of extreme earthquake	1000 years	2500 years
Modeling: Soil-pile-Structure Interaction Modeling parameters and Criteria for members	By details Don't have	Global Have
Simplified procedure for assessment	---	Coefficient Method
Rehabilitation method	Don't have	Have

Table 2. Comparison of API RP 2A and FEMA-356

1. Comparing with buildings, the redundancy of the jacket platform structure is low, and failure of some members can affect not only the routine operation of structure such as production activities, but also the safety and serviceability of the structure. Then the results of FEMA are more reliable and economical than API.
2. Specific criteria can be taken into account for the evaluation of each member with respect to its condition. This matter is more important for a jacket platform in that strength degradation of members due to fatigue and corrosion and etc. is feasible.
3. Rehabilitation of the structure can be performed better knowing the behavior of members.

The return period adopted for collapse requirement ground motions in API RP 2A is lower than that of FEMA-356. Earthquake return periods of FEMA and API are compared to each other in the Table 4. From a point of comparison with FEMA-356, there are four principal reasons why the earthquake return period of API is low:

1. Importance of offshore structures is higher than buildings.
2. Seismic loads imposed on a structure are highly dependent on the stiffness and energy dissipation characteristics of the structural system, including the piling and supporting soils, and so higher uncertainties in soil properties result directly in higher uncertainties in loads.
3. Uncertainties in the estimation of ground motions for offshore structures are higher than those for buildings.
4. Because of lower redundancy, the sensitivity to increase in return period is greater for offshore structures than buildings. This item is an important difference between jacket platforms and buildings.

The FEMA document represents simplified procedures such as a coefficient method that can be used for assessment and rehabilitation of buildings for different loading conditions and this document consists of useful procedures for the rehabilitation of buildings.

2.4 Response determination using nonlinear pushover analysis

Nonlinear time-history analysis can be used for assessment and rehabilitation of all types of structures. This procedure is complicated and time consuming. Nowadays nonlinear static procedures are widely used for the assessment and rehabilitation of structures. These procedures can be used to estimate the response of structures under seismic loading. The target displacement for each level of load is calculated. The target displacement is intended to represent the maximum displacement likely to be experienced during the design earthquake. The stresses and deformations in each component are then evaluated at this displacement level. FEMA-356 utilizes the Coefficient Method in which several empirically derived factors are used to modify the response of a single-degree-of-freedom (SDOF) model of the structure assuming that it remains elastic. The Capacity-Spectrum Method of ATC-40 uses empirically derived relationships for the effective period and damping as a function of ductility to estimate the response of an equivalent linear SDOF oscillator. Recently these methods evaluated and improved in the FEMA-440 document [16].

2.4.1 Capacity-spectrum method (ATC-40)

In the Capacity-Spectrum Method, the base shear versus roof displacement relationship (capacity) and seismic ground motion (demand) are plotted in Acceleration-Displacement Response Spectrum (ADRS) format. The performance point (maximum inelastic displacement) can be obtained from the intersection point of demand and capacity. This procedure is presented in Fig. 3. In this figure S_a and S_d are spectral acceleration and spectral displacement respectively.

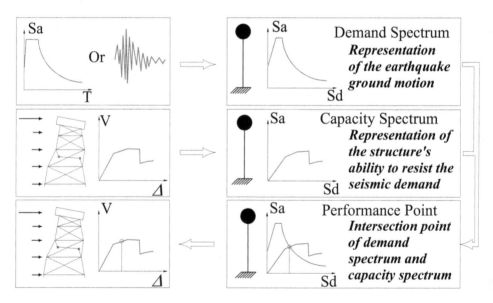

Fig. 3. Capacity-spectrum method.

2.4.2 The coefficient method (FEMA-356)

In the Coefficient Method, the maximum inelastic displacement (Target Displacement) is obtained from multiplying the linear elastic response by a series of coefficients C_0 through C_3. These coefficients are derived from statistical studies of the nonlinear time-history analyses of SDOF oscillators.

2.4.3 FEMA-440 recommendations

The FEMA-440 document evaluates and improves the abovementioned simplified inelastic analysis procedures. Proposed modifications to the Coefficient Method of FEMA-356 relate primarily to the coefficients themselves. For coefficients C_1 and C_2 new relationships are proposed. It is also proposed that instead of coefficient C_3 a limitation on minimum strength be used. The improved procedure for the Capacity-Spectrum Method consists of new estimates of equivalent period and damping. This Linearization Method is calibrated for certain hysteretic loops with different calibration equations for the nondegrading and degrading cases.

2.5 Case study
2.5.1 Description of the jacket platform

The existing offshore complex consists of a drilling platform, a production platform, a service platform and a flare tripod in the field. The field was originally developed and put in production in 1968. There has been some damage imposed during Iran/Iraq war and some other extended damage due to adverse climate conditions afterwards. The service platform consists of a four leg battered jacket and topside located in 67.40 m water depth which is connected to production platform by means of one existing bridge. The service life of the platform is 25 years.

2.5.2 Load cases

For time history analysis of the platform, a `best fit' set of scaled, natural time histories is used provided the velocity spectrum values have been properly modulated to equal or exceed the standard spectrum velocity values at specified periods (0.2 T to 1.5 T) as mentioned in International Building Code (IBC).

2.5.3 Numerical model

Analytical models were created using the open source finite element platform, OpenSees. This program is useful for modeling of jacket platform structures because of its capability of modelling of the post-buckling behavior of tubular members, soil-pile-structure interaction and etc. A two-dimensional model of a single frame is developed for the structure. A force-based nonlinear beam-column element (utilizing a layered fiber section) is used to model all components of the frame. Steel material is modeled using a bilinear stress-strain curve with 0.3% post-yield hardening. Initial imperfections in the struts are accounted for, with a value of 0:001L where L is the length of the member. This idea is useful for modeling the post-buckling behavior of the strut members, respectively.

The mathematical model of the pile-soil-structure system consists of the following sets of elements (Fig. 4):

1. Pile elements, modeled by a number of nonlinear beam-column elements.
2. Far-field soil model representing the free-field motion of the soil column, vertically and horizontally that is unaffected by the pile motions. The soil is modeled using elastic quad elements. The nodes that are at the same depth are constrained.
3. Near-field elements that connect the piles to the soil, vertically and horizontally. The strength and stiffness of these elements depends on the state of the far-field soil and the relative motion of the pile and far-field soil. The interface between the pile and

surrounding soil is modeled using p-y, t-z, q-z nonlinear spring elements. Hysteretic and radiation damping are considered using these elements. The group effects are not considered. The input motion is applied to the fixed nodes at the bottom of the soil column. The seismic record at bedrock is found from the input motion at the surface. Hydrodynamic effects are considered in terms of hydrodynamic damping from drag forces and added masses.

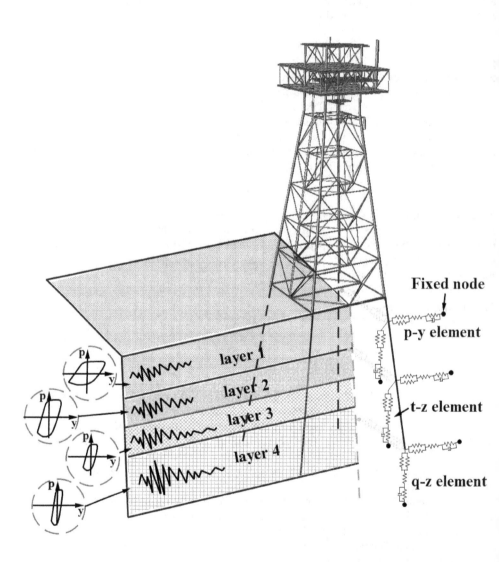

Fig. 4. Modeling of soil-pile-structure interaction.

2.5.4 Numerical study

A uniform distribution of load is applied for the pushover analysis. The backbone curve of the deformation-control action of members is needed for an assessment of the structure based on the FEMA procedure. Fig. 5 shows the backbone curve of the axial action for strut-2 of the platform and FEMA criteria. Table 6 shows dimensions and assessment parameters of jacket struts. In this Table Δ_C and Δ_T are the axial deformations at the expected buckling load and at the expected tensile yielding load, respectively. Fig. 6 shows the pushover curve of the structure. This Figure shows an instantaneous loss of strength at a deck displacement equal to 0.29 m. It can be seen that after the point with a deck displacement of 1.15m the load-deformation curve has a negative slope. This figure also shows that for a deck displacement equal to 2.0 m, the 1st and 2nd platform levels remain elastic.

The typical platform in Persian Gulf was assessed using responses obtained from a series of nonlinear time history analyses using three best fit records for each hazard level.

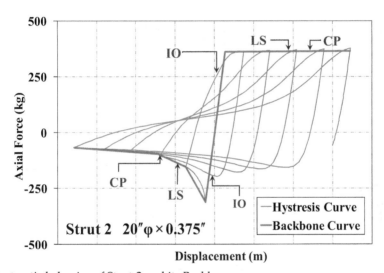

Fig. 5. Hysteretic behavior of Strut-2 and its Backbone curve.

2.5.5 Structural assessment

This section provides processes of assessment of the typical jacket platform based on the API and FEMA-356. The demands are obtained from mean values of the time history analysis.

2.5.5.1 API standard

The collapse of the structure is defined by its lack of ability to withstand to the load. The collapse load is defined as the maximum load the structure can withstand, before the load-displacement curve starts a negative trend. The deck displacement at the collapse point of the jacket is 1.15 m.

2.5.5.2 FEMA-356 document

Leg behavior. Compression action in leg elements is a force control action, but the action type for flexure in these elements depends on the axial force of legs.

Brace behavior. Braces are members with deformation controlled action and mainly determine the performance level of jackets. Fig. 6 shows the performance of the jacket for each hazard level. This figure shows that the jacket platform is highly sensitive to input motion.

2.5.6 Directivity effects

One of the primary factors affecting motion in the near-fault region is the directivity in which rupture progresses from the hypocenter along the zone of rupture. "Directivity" refers to the direction of rupture propagation as opposed to the direction of ground displacement (Abrahamson, 1998). A site may be classified after an earthquake as demonstrating forward, reverse, or neutral directivity effects. If the rupture propagates toward the site and the angle between the fault and the direction from the hypocenter to the site is reasonably small, the site is likely to demonstrate forward directivity. If rupture propagates away from the site, it will likely demonstrate reverse directivity (Abrahamson, 1998). If the site is more or less perpendicular to the fault from the hypocenter it will likely demonstrate neutral directivity. The phrase "directivity effects" usually refers to "forward directivity effects", as these case results in ground motions that are more critical to engineered structures.

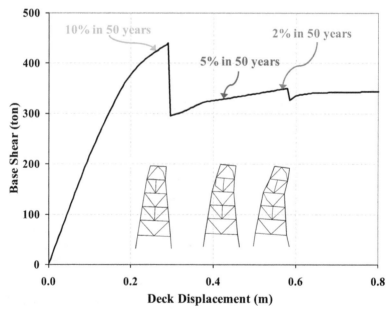

Fig. 6. Performance of the jacket for various levels of hazard.

A large velocity directivity pulse occurs when the conditions of forward directivity are met. These conditions include:
1.   The earthquake is sufficiently large (moment magnitude greater than 6);
2.   The site is located sufficiently close to the fault rupture (within 10 km); and
3.   The rupture propagates toward the site.

This large velocity directivity pulse will be evident in the fault-normal direction. It is typically located toward the front of the time history and consists of, on average, one cycle of motion. Peak velocities usually are between 30 cm/sec to 200 cm/s with a mean value about 100 cm/sec. The period of the pulse can range from 0.5 sec to 5 sec with a mean value about 2.5 sec. Tabeshpour presented a conceptual discussion on the effect of near field earthquakes on various structures (2009).

2.5.6.1 Pulse-type excitation

For consideration of directivity effects, the analytical velocity pulse model proposed by He (2003) is expressed as:

$$\dot{u}_p = Ct^n e^{-at} \sin \omega_p t \tag{1}$$

Where C is the amplitude scaling factor, a is the decay factor, n is the shape parameter of the envelope, and ω_p is the pulse frequency in rad/s. Differentiating above equation, the acceleration $\ddot{u}_p$ of the pulse can be obtained as:

$$\ddot{u}_p = Ct^n e^{-at} \left[(\frac{n}{t} - a) \sin \omega_p t + \omega_p \cos \omega_p t \right] \tag{2}$$

The acceleration $\ddot{u}_p$ in above Equation is considered as ground acceleration for numerical simulations in this chapter. To illustrate the performance of structure during the near-fault excitations, the parameter $\beta = T_p/T$ is used, where T and T_p are the fundamental period of jacket platform and pulse period, respectively. Fig. 7 shows time-history plots of acceleration for a velocity pulse with parameters $a = 2.51, C = 7.17, n = 1$ for three T_p.

2.5.6.2 Results and discussions

Base shear of the structure vs. deck displacement (hysteresis loops) is presented in Fig. 8. A global view of the nonlinear behavior of the structure is seen clearly at the first pulse of excitation. In is seen that all energy dissipation by hysteretic behavior of the elements is occurred just in the one loop. Maximum displacements are shown by points A, B and C in horizontal axis.

3. Rehabilitation

Fig. 9 provides rehabilitation process of jacket offshore platforms. Rehabilitation of existing jacket consists of two phases of assessment and rehabilitation. Many researches have been carried out in this matter that is shown in this figure. With respect to type of loads, many types of control systems can be used for rehabilitation of jacket offshore platforms. Fig. 10 shows types of control systems usable for jacket offshore platforms.

3.1 Tuned mass damper (TMD), wind and wave protection

TMD is suitable passive control device for narrow band loads such as wind and wave loads. The efficiency of this device should be investigated for environmental loads. Fig. 11 shows jacket platform equipped with TMD and equal single degree of freedom. Fig. 12 shows the effect of TMD on the displacement response of jacket platform under the harmonic wave load with wave period equal to the fundamental period of the structure (T) and wave height H=0.212 m. A clearly decrease is observed in structural response.

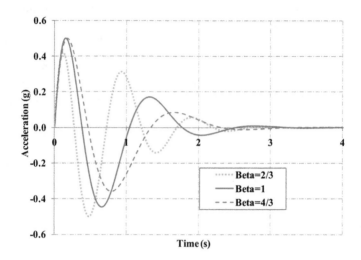

Fig. 7. Time-history plots of acceleration for $\beta = 2/3, 1, 4/3$.

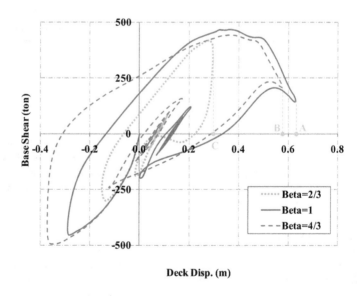

Fig. 8. Hysteresis loops of the structure for $\beta = 2/3, 1, 4/3$.

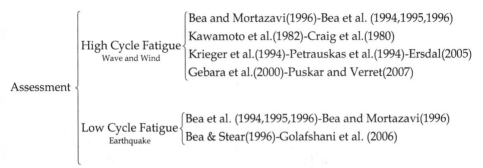

Fig. 9. Rehabilitation studies process for jacket offshore platforms.

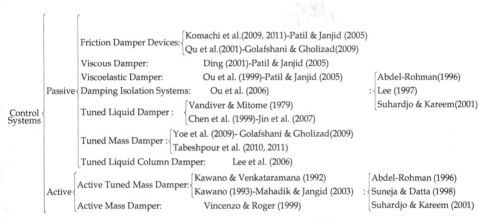

Fig. 10. Types of control systems used for jacket offshore platforms.

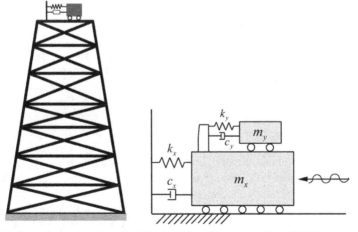

Fig. 11. Steel jacket platform utilized with a TMD and its equivalent SDOF system.

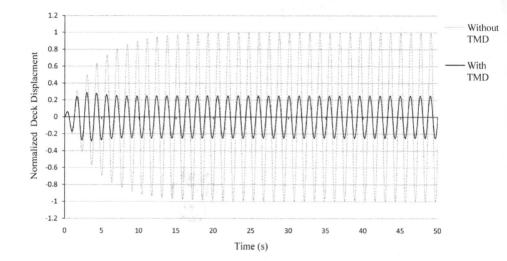

Fig. 12. Time history of deck displacement for harmonic load; W/O & W TMD (H=0.212 m, T).

Fig. 13 also shows a clearly decrease in structural response for the harmonic wave load with wave period equal to three times of the fundamental period of the structure (3T) and wave height H=1.91 m. Time history of deck acceleration for this loading has been shown in Fig. 14. It can be seen that acceleration reduces highly.

Fig. 15 shows the effect of TMD on the response of structure opposed to wind load. Mean velocity of wind with return period of 50 years at 10 m height from sea surface has been assumed to be equal to 22.5 m/s. This figure shows that TMD is effective for wind load cases as well.

3.2 Friction damper devices (FDD), seismic protection

In this research, a novel friction damper device (FDD), Mualla and Belev (2002), which is economical, can be easily manufactured and quickly installed, is used. The damper main parts are the central (vertical) plate, two side (horizontal) plates and two circular friction pad discs placed in between the steel (Fig. 16). The hinge connection is meant to increase the amount of relative rotation between the central and side plates, which in turn enhances the energy dissipation in the system. The ends of the two side plates are connected to the members of inverted V-brace at a distance r from the FDD centre. The bracing makes use of pretensioned bars in order to avoid compression stresses and subsequent buckling. The bracing bars are pin-connected at both ends to the damper and to the column bases. The combination of two side plates and one central plate increases the frictional surface area and provides symmetry needed for obtaining plane action of the device. Zero-length element of program used for modeling of the frictional hinge. In order to verify modeling assumptions, model related to Mualla article evaluated. Tabeshpour and Ebrahimian presented a simple procedure for design of friction damper (2007, 2009). Komachi et al. investigated the efficiency of FDD for rehabilitation of jacket platforms (2011).

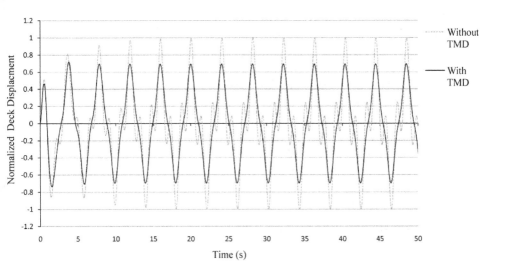

Fig. 13. Time history of deck displacement for harmonic load; W/O & W TMD (H=1.91 m, 3T).

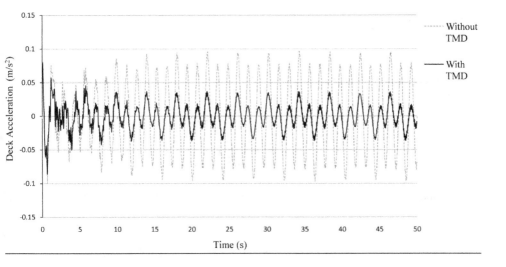

Fig. 14. Time history of deck acceleration for harmonic load; W/O & W TMD (H=1.91 m).

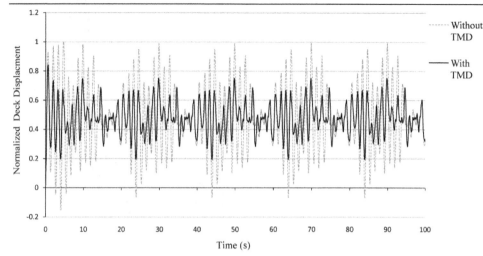

Fig. 15. Time history of deck displacement for wind load; W/O & W TMD.

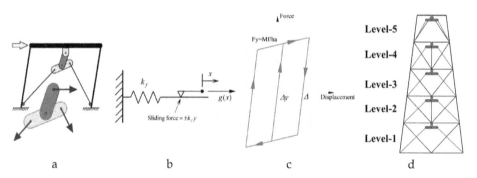

Fig. 16. a) Configuration of friction damper, b) Mathematical model, c)Hystresis behavior, d) Arrangement of FDD.

The equivalent viscous damping of this system is obtained by:

$$\beta_{eff} = \frac{2}{\pi} \frac{FR(SR - FR)}{\left(SR + FR^2\right)}, \quad \frac{FR}{SR} \prec 1 \tag{3}$$

That FR and SR are the damper properties in terms of the structure properties are defined as follows:

$$\begin{cases} FR = \dfrac{F_y}{F_e}: & \text{the ratio of damper stiffness to total structure stiffness} \\[3mm] SR = \dfrac{K_{bd}}{K_e}: & \text{the ratio of damper yield force to total structure force} \end{cases}$$

Fig. 17 shows the comparison of pushover curves of the jacket with and without the damper. It can be seen that the damper improves the performance of the jacket especially at the nonlinear region. Base shear of the structure vs. deck displacement (hysteresis loops) for CHY101W record is presented in Fig. 18 for cases of with and without FDD.

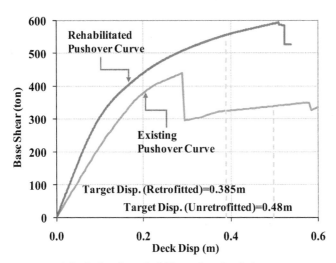

Fig. 17. Pushover curve of the jacket for rehabilitated and existing cases.

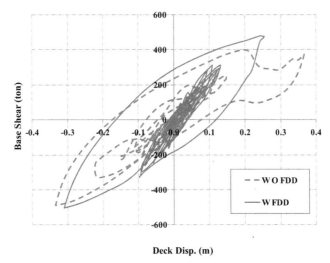

Fig. 18. Hysteresis loops of jacket for CHY101W record.

Fig. 19 shows time history of frictional hinge rotation of dampers at various levels of jacket for CHY101W record. Fig. 20 shows the time history of deck displacement for CHY101W record. This figure shows that deck displacement reduces highly (about 60%) and base shear of structure reduce about 10% too.

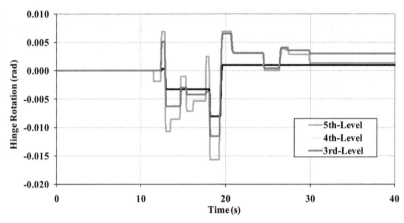

Fig. 19. Time history of friction hinge rotation for CHY101W record.

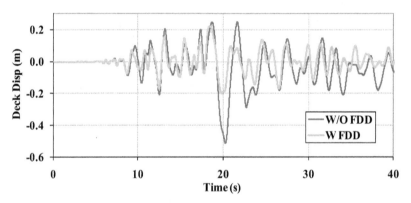

Fig. 20. Time history of deck displacement of structure for cases with and without damper for CHY101W record.

4. Conclusion

This chapter presented process of assessment and rehabilitation of jacket platform offshore structures. A seismic assessment of API standard and FEMA-356 documents was compared to each other. As an example an existing 4 legged service platform was assessed with API and FEMA-356 with respect to various earthquake hazard levels. Nonlinear static and dynamic analyses were used to determine the response of the structure. This study shows that:

1. Comparative building redundancy of a jacket platform structure is low and failure of one member in these structures can lead to immediate reduction of strength and afterward collapse of the structure.
2. However global criteria for seismic assessment of jacket platforms are presented in the API standard but there is no numerical or specific criterion in order to assess the structure. It is observed that building documents can be used to develop numerical and

applicable criteria for seismic assessment of such structures. Assessment with respect to member's criteria has some advantages such as better decision making for rehabilitation and detecting the full capacity of members.

3. Return periods related to collapse prevention in API and FEMA are 1000 and 2500 years respectively. The return period of API should probably be reviewed because the expected mean life time of the jacket is greater than design mean life time. However the approach and methodology presented in building structure documents (such as FEMA-356) is very appropriate and efficient in the seismic assessment of jacket platform structures.

The effects of near-fault earthquakes on the behavior of steel jacket platforms has been presented. Pulse type excitation has been used for investigation of structural behavior. It is shown that the maximum response of structure occurs when $T = T_p$ and also that input pulse with $\beta > 1.0$ gives a higher amplitude rather than $\beta < 1.0$, that denotes importance of ratio of period of directivity pulse to structure period. The increasing in dynamic amplitude can be more than two times than that of both static and far field responses.

Rehabilitation process using TMD and FDD for these structures presented. Effect of TMD on the response of jacket offshore platform under the wave and wind loads presented. It was shown that TMD is very effective for reduction of jacket responses under the these loads.

FDD was used on a steel jacket platform located in Iranian waters of the Persian Gulf. Results were shown that responses of jacket reduce dramatically. A numerical study was performed using pushover and nonlinear dynamic analysis. Pushover analysis results were shown that use of FDD system reduce target displacement of the structure and also was shown that a sudden decrease of jacket strength does not occur when this system is installed on the structure. Due to the low redundancy of jacket platform structures, the strength of these structures can decrease suddenly and the use of FDD systems can be extremely useful. Analysis results were shown that friction damper greatly reduces deck displacement. It was observed that for large record accelerations structure behavior becomes highly nonlinear and the performance of the friction damper for response reduction increases (for example up to 65% deck displacement reductions). Numerical studies clearly exhibit that these control systems represent a practical alternative for rehabilitation of existing jacket platforms.

5. References

Abdel-Rohman, M. (1996). Structural control of steel jacket platform. *Structural Engineering and Mechanics*, 4, 25–38

American petroleum institute recommended practice for planning, design and constructing fixed offshore platforms. API RP 2A, 21th ed. Washington (DC): American Petroleum Institute; 2000

Anagnostopoulos, S. (October 1978). Post-yield Flexural Properties of Tubular Members. *Preprint 3302, ASCE Convention and Exposition*, Chicago

Applied Technology Council (ATC). Seismic evaluation and retrofit of concrete buildings. vol. 1 and 2. Report no. ATC-40. Redwood City (CA): 1996

Bea, R.G.; Loch, K. J. & Young, P. L. (October 2, 1994). Screening Methodologies for Use in Platform Assessments and Requalifications. Department of Civil Engineering, University of California at Berkeley

Bea, R.G.; et. al. (May 1-4, 1995). Verification of a Simplified Methods to Evaluate the Capacities of Template-Type Platforms, *27th Annual Offshore Technology Conference*, Houston, Texas

Bea, R. G.; Mortazavi, M. (May 26-31, 1996). A Simplified Structural Reliability Analysis Procedure for Use in Assessments and Requalifications of Template-Type Offshore Platforms, *6th International Offshore and Polar Engineering Conference*, Los Angeles, California

Bea, R.G.; Stear, J.D. (March 1996). Using Static Pushover Analysis to Determine the Ultimate Limit States of Gulf of Mexico Steel Template-Type Platforms Subjected to Hurricane Wind and Wave Loads. *Marine Technology and Management Group Project*, Department of Civil Engineering, University of California at Berkeley

Bea, R.; Stear, J. (March 1999). Report 3: Benchmark Platform and Site Studies, Department of Civil & Engineering, University of California at Berkeley

Briaud, J. L.; Tucker, L. M. & Lytton, R. L. & Coyle, H. M. (September 1983). Behavior of Piles and Pile Groups in Cohesionless Soils, Texas A&M University, College Station, Texas

Briaud, J. L.; Tucker, L. M. (October 1983). Piles in Sand: A Design Method Including Residual Stresses, Texas A&M University, College Station, Texas

Briaud, J. L.; Tucker, L. M. (May 17, 1984). Coefficient of Variation of In-Situ Tests in Sand, ASCE Symposium `Probabilistic Characterization of Soil Properties: Bridge Between Theory and Practice, Atlanta, Georgia

Bardakis, V.G.; Dritsos, S.E. (2007). Evaluating assumptions for seismic assessment of existing buildings. *Soil Dyn Earthq Eng*, 27(3), 223_33

Chen, X.; Wang, L.Y. & Xu, J. (1999). TLD technique for reducing ice-induced vibration on platforms. *J Cold Reg. Eng*, 13(3), 139–52

Cornell, C. A. (August 1987). Offshore Structural Systems Reliability, C. Allin Cornell, Inc., Portola Valley, California

Craig, M.J.K.; Shekher, V. (1980). Inelastic Earthquake Analysis of an Offshore California Platform, *Proceedings, Offshore Technology Conference*, OTC 3822, Houston, Texas, 259−267

Craig, M.J.K.; & Digre, K.A. (1994). Assessment of high-consequence platforms; Issues and applications, *Proceedings of offshore technology conference*, OTC 7485, Houston (TX)

Ding, J.H. (2001). Theoretical and experimental study on structural vibration repressed system using viscous fluid dampers, *Ph.D. dissertation*, Harbin Institute of Technology, [in Chinese]

El Naggar, M.H.; Novak, M. (1996). Influence of foundation nonlinearity on offshore towers response. *J Geotech Engng*, ASCE, 1996, 122(9), 717–24

Elnashai, A. S.; Aritenang, W. (1991). Nonlinear modelling of weld-beaded composite tubular connections. *Eng. Struct*, 1991, Vol 13 34-42

Ettema, R.; Stern, F. & Lazaro, J. (April 1987). Dynamics of Continuous-Mode Icebreaking by a Polar-Class Icebreaking Hull. *IIHR Report*, Iowa Institute of Hydraulic Research, The University of Iowa, Iowa City, Iowa

Ersdal, G. (2005). G. Assessment of existing offshore structures for life extension, *Ph.D. dissertation*, University of Stavanger

FEMA-356 (2000). Prestandard and commentary for the seismic rehabilitation of buildings. Report FEMA-356. Washington (DC): Federal Emergency Management Agency

FEMA-440 (2005). Improvement of nonlinear static seismic analysis procedures. Report FEMA-356. Washington (DC): Federal Emergency Management Agency

Ferritto, J. M.; Nakamoto, R. T. (September 1982). The Effective Stress Soil Model, Technical Memorandum 51-82-07, Naval Civil Engineering Laboratory, Port Hueneme, California

Filiatrault, A.; Cherry, S. (1989). Parameters influencing the design of friction damped structures. Department of civil eng., University of British Columbia, Vancouver, B.C., Canada V6T 1W5, 753-766.

Gebara, J.; Dolan, D. & Pawsey, S. & Jeanjean, P. & Dahl-Stamnes, K. (2000). Assessment of offshore platforms under subsidence Part I: Approach. *ASME J Offshore Mech Arct Eng*, 122, 260-6

Golafhshani, A.A.; Tabeshpure, M.R. & Komachi, Y. (2006). Assessment and retrofit of existing jacket platform using friction damper devices, *8th international conference of ocean industries*, Bushehr (Iran), Bushehr University, [in Persian]

Golafhshani, A.A.; Gholizad, A. (2009). Friction damper for vibration control in offshore steel jacket platforms. *Journal of Constructional Steel Research*, 65(1), 180-187

Golafhshani, A.A.; Tabeshpure, M.R. & Komachi, Y. (2009). FEMA approaches in seismic assessment of jacket platforms, (case study: Ressalat jacket of Persian gulf). *Journal of Constructional Steel Research*, 65(10), 1979-1986

Grilli, S.T.; Fake, T. & Spaulding, M.L. (May 26, 2000). Numerical Modeling of Oil Containment by a Boom/Barrier System: Phase III Final Report. University of Rhode Island

Hueste, M.D.; Bai, J.W. (2007). Seismic retrofit of a reinforced concrete flat-slab structure: Part I seismic performance evaluation. *Eng Struct*, 29(6), 1165-77

IBC, 2000. International Building Code. International, Code Council. Falls Church (Virginia).

ISO (2000). ISO/DIS 13822 Bases for design of structures_Assessment of existing structures. International Standardisation Organisation

ISO (2002). ISO 19900 Petroleum and natural gas industries_Offshore structures_Part 1: General requirements. International Standardisation Organisation

ISO (2004). ISO/DIS 19902 Design of fixed steel jackets. DIS Draft. International Standardisation Organisation

Jin, Q.; Li, X. & Sun, N. & Zhou, J. & Guan, J. (2007). Experimental and numerical study on tuned liquid dampers for controlling earthquake response of jacket offshore platform. *Marine Structures*, 20(4), 238-254

Kallaby, J.; Millman, D.N. (1975). Inelastic Analysis of Fixed Offshore Platforms for Earthquake Loading, *Proceedings, Offshore Technology Conference*, Houston, Texas

Kawamoto, J.; Sunder, S. and Connor, J. (1982). An assessment of uncertainties in fatigue analyses of steel jacket offshore platforms. *Applied Ocean Research*, Vol. 4, No.1, 9-16

Kawano, K.; Venkataramana, K. (1992). Seismic response of offshore platform with TMD, *Proc. of 10th World Conference on Earthquake Engineering*, 4, 241–246

Kawano, K. (1993).Active control effects on dynamic response of offshore structures, *Proc. of 3rd ISOPE Conference*, 3, 494–498

Kawano, K.; Venkataramana, K. (January 1999). Dynamic response and reliability analysis of large offshore structures, *Computer Methods in Applied Mechanics and Engineering*, Volume 168, Issues 1-4, 6, 255-272.

Komachi, Y.; Tabeshpour, M.R. & Golafhshani, A.A. (2009a).Assessment of seismic vulnerability of jacket platforms based on the performance, *11th international conference of ocean industries*, Kish, Iran, (in Persian)

Komachi, Y.; Tabeshpour, M.R. & Golafhshani, A.A. (2009b).Use of energy damping systems for rehabilitation of jacket platforms, *11th international conference of ocean industries*, Kish, Iran, [in Persian]

Komachi, Y.; Tabeshpour, M.R. & Golafhshani, A.A. (2010). Special considerations for finite element modeling of jacket platforms, *12th international conference of ocean industries*, Zibakenar, Iran, [in Persian]

Komachi, Y.; Tabeshpour, M.R. & Mualla, I. & Golafhshani, A.A. (2011). Retrofit of Ressalat Jacket Platform (Persian Gulf) using Friction Damper Device. *Journal of Zhejiang University*, Sci A (Appl Phys & Eng) 2011 12(9):680-691

Krieger, W.F.; Banon, H. & Lloyd, J.R. & De, R.S. & Digre, K.A. & Nair, D. et al. (1994). Process for assessment of existing platforms to determine their fitness for purpose, *Proceedings of offshore technology conference*, OTC 7482, Houston (TX)

Lee, H.H. (1997). Stochastic analysis for offshore structures with added mechanical dampers. *Ocean Engineering*, 24, 817–834

Lee, H.N.; Wong, S.H. & Lee, R.S. (2006). Response mitigation on the offshore floating platform system with tuned liquid column damper. *Ocean Eng.*, 33(8–9), 1118–42

Lie, S.T.; Lee, C.K. & Chiew, S.P. & Shao, Y.B. (2005). Validation of surface crack stress intensity factors of a tubular K-joint. *International Journal of Pressure Vessels and Piping*, 82, 610–617

Litton, R.W.; Pawsey, S.F. & Stock, D. & Wilson, B.M. (1978). Efficient Numerical Procedures for Nonlinear Seismic Response Analysis of Braced Tubular Structures. *Preprint 3302, ASCE Convention and Exposition*, Chicago, October

Mahadik, A.S.; Jangid, R.S. (2003). Active control of offshore jacket platforms. *International Shipbuilding Progress*, 50, 277–295

Maison, B.; Popov, E.P. (1980). Cyclic response prediction for braced steel frames. *J. Struct., Div.*, ASCE, 106, 1401 – 1416

Marshall, P.W.; Gates, W.E. & Anagnostopoulos, S. (1977). Inelastic Dynamic Analysis of Tubular Offshore Structures, *Proceedings, Offshore Technology Conference*, OTC 2908, Houston, Texas, 235 – 246

Mitwally, H.; Novak, M. (1987). Response of offshore towers with pile interaction. *J Geotech Engng*, ASCE, 1987, 113(7), 1065–84

Moustafa, A. (March, 2011). Damage-Based Design Earthquake Loads for Single-Degree-Of-Freedom Inelastic Structures, *Journal of structural engineering*, Vol. 137, No. 3, 456-467

Mualla, I.H.; Belev, B. (2002). Performance of steel frames with a new friction damper device under earthquake excitation. *Engineering Structures*, 24, 365-371

Mualla, I.H.; Nielsen, L.O. (2002). A Friction Damping System Low order behavior and design. Department of Civil Engineering, DTU-bygning, 118

Nath, J. H.; Hsu, M. K. & Hudspeth, R.T. & Dummen, J. (September 11-13, 1984). Laboratory Wave Forces on Vertical Cylinders, *Proceedings Ocean Structural Dynamics Symposium*, Oregon State University, Corvallis, Oregon, pp. 312-330

Nogami, T.; Konagai, K. (February 1987). Dynamic Response of Vertically Loaded Nonlinear Pile Foundations. Paper No. 21252, *ASCE Journal of Geotechnical Engineering*, Volume 113, No. 2

Nogami, T.; Leung, M. B. (March 1-6, 1987). Dynamic Soil-Structure Analysis of Offshore Gravity Platforms Using Simplified Model, *Proceedings of the Sixth International Offshore Mechanics and Arctic Symposium*, Houston, Texas

OpenSees 2005. Open system for earthquake engineering simulation. Available online http://opensees.berkeley.edu.

Ou, J.P.; Duan, Z.D. & Xiao, Y.Q. (1999). Ice-induced vibration analysis of JZ20-2MUQ offshore platform using in-situ ice force histories. *The Ocean Engineering*, 17(2), 70-80

Ou, J.P.; Xiao, Y.Q. & Duan, Z.D. & Zou, X.Y. & Wu, B. & Wei, J.S. (2000). Ice-induced vibration control of JZ20-2MUQ platform structure with viscoelastic energy dissipators. *The Ocean Engineering*, 18(3), 9-14

Ou, J.; Long, X. & Li, Q.S. & Xiao, Y.Q. (2006). Vibration control of steel jacket offshore platform structures with damping isolation systems. *Engineering Structure*, 29(7), 1525-1538

Patil, K.C.; Jangid, R.S. (2005). Passive control of offshore jacket platforms. *Ocean Engineering*, 32, 1933-1949

Petrauskas, C.; Finnigan, T.D. & Heideman, J.C. & Vogel, M. & *Santala*, M. & Berek, G.P. (1994). Metocean criteria/loads for use in assessment of existing offshore platforms, *Proceedings of offshore technology conference*, OTC 7484, Houston (TX)

PSA (2004). Regulation relating to management in the petroleum activities. Petroleum Safety Authority Norway. 2004

Puskar, F.J.; Verret, S.M. and Roberts, C. (2007). Fixed platform performance during recent hurricanes: comparison to design standards. *OTC 18989*, Presented at the 2007 Offshore Technology Conference. Houston, TX

Qu, W.L.; Chen, Z.H. & Xu, Y.L. (2001). Dynamic analysis of wind-excited truss tower with friction dampers. *Computers & Structures*, 79(32), 2817-2831

Ressalat (R1). Offshore complex renovation and reconstruction project. Production platform in place analysis report. Rsl-P1-St-Cn-1001-C2. Amid Engineering & Development Company

Soong, T.T.; Dargush, G.F. (1997). *Passive energy dissipation systems in structural engineering*. Wiley, London

Suhardjo, J.; Kareem, A. (2001). Feedback–feedforward control of offshore platforms under random waves. *Earthquake Engineering and Structural Dynamics*, 30(2), 213-235

Suneja, B.P.; Datta, T.K. (1998). Active control of ALP with improved performance functions. *Ocean Engineering*, 25, 817-835

Suneja, B.P.; Datta, T.K. (1999). Nonlinear open-close loop active control of articulated leg platform. *International Journal of Offshore and Polar Engineering*, 9, 141-148

Tabeshpour, M.R.; Ebrahimian, H. (2007). Application of friction dampers in seismic retrofit of existing structures, *5th international conference on seismology and earthquake engineering*, SEE5, Tehran, Iran

Tabeshpour, M.R. (2009). Effect of near field motions on seismic behavior of long period structures. *Bana Journal*, Vol. 11, No.1 [In Persian]

Tabeshpour, M.R.; Ebrahimian, H. (2010). Seismic Retrofit of Existing Structures using Friction. *Asia Journal of Civil Engineering*, 11(4), 509-520

Tabeshpour, M.R.; Dehkharghanian, V. & Dovlatshahi, M. (2010). Effect tuned mass damper on vertical vibration of tension leg platform, *MIC2010*, Zibakenar, Iran, [In Persian]

Tabeshpour, M.R.; Rezaie, E. & Arafati, N. (2011). Response mitigation of jacket platforms using tuned mass damper, *1st National conference on steel structures*, Tehran, Iran

Takewaki, I.; Murakami, S. & Fujita, K. & Yoshitomi, S. & Tsuji, M. (2011). The 2011 off the Pacific coast of Tohoku earthquake and response of high-rise buildings under long-period ground motions, *Soil Dynamics and Earthquake Engineering*, In Press

Terro, M.J.; Mahmoud, M.S. & Abdel-Rohman M. (1999). Multi-loop feedback control of offshore steel jacket platforms. *Computers & Structures*, 70(2), 185-202

Toma, S.; Chen, W.F. (1982). Inelastic cyclic analysis of pin-ended tubes. *J Struct Div.*, ASCE, 108, 2279 − 2294

Vandiver, J.K.; Mitome, S. (1979). Effect of liquid storage tanks on the dynamics response of offshore platform. *Applied Ocean Research*, 1, 67–74

Veletsos, A.S. (1977). Dynamics of Structure-Foundation Systems. *Structural and Geotechnical Mechanics*: A Volume Honoring Nathan M. Newmark, editor: W. J. Hall, Prentice-Hall, Englewood Cliffs, NJ

Venkataramana, K.; Kawano, K. & Yoshihara, S. (May 24-29, 1998). Time-domain dynamic analysis of offshore structures under combined wave and earthquake loadings, *proceeding of the eighth international offshore and polar engineering conference*, Montreal, Canada

Vincenzo, G.; Roger, G. (1999). Adaptive control of flow-induced oscillation including vortex effects. *International Journal of Non-Linear Mechanics*, 34, 853–68

Wang, S. (2002). Semi-active control of wave-induced vibration for offshore platforms by use of MR damper, *International Conference on Offshore Mechanics and Artic Engineering*, Oslo, Norway, 23-28

Xu, T.; Bea, R. (January 1998). Reassessment of the Tubular Joint Capacity: Uncertainty and Reliability. *Screening Methodologies Project Phase IV, Report to Joint Industry Project Sponsors, Marine Technology and Management Group*, Department of Civil Engineering, University of California at Berkeley

Yue, Q.; Zhang, L. & Zhang, W. & Kärnä, T. (2009). Mitigating ice-induced jacket platform vibrations utilizing a TMD system. *Cold Regions Science and Technology*, 56(2-3), 84-89

Zayas, V.A.; Popov, E.P. & Mahin, S.A. (1980). Cyclic inelastic buckling of tubular steel braces. *Report UCB/EERC-80/16*, Earthquake Engineering Research Center, Berkeley (CA), University of California

6

Electromagnetic Sensing Techniques for Non-Destructive Diagnosis of Civil Engineering Structures

Massimo Bavusi et al.*
CNR-IMAA,
Italy

1. Introduction

Health Assessment Methods (HAM) and Structural Health Monitoring (SHM) aim to improve the standard of knowledge regarding the safety and maintenance of structures and infrastructure acquiring information about geometrical, mechanical and dynamical characteristics of structures. In earthquake-prone areas, this activity has the double aim of assessing the buildings structural integrity and extracting information regarding their response during a seismic event in order to define appropriate activities for risk mitigation.

A number of factors afflict buildings and infrastructure safety in seismic areas:

- Outdated codes of practice: a significant number of highly urbanized areas are present globally, where a high percentage of structures have been designed and erected considering only gravity loading.
- The age of the structures and the real in-situ performance of construction material significantly affect their overall behaviour.
- Structural deficiencies such as poor material qualities and/or degradation of structural materials (rust, spalling etc.), inadequate construction detailing, low levels of ductility, brittle collapse mechanisms.

The seismic assessment of structures is performed in terms of the estimation of the earthquake intensity that would lead to a certain damage condition and/or collapse. The assessment of the seismic vulnerability of existing buildings is generally based on the knowledge of building characteristics and through a complex analysis of the possible collapse mechanisms in order to identify the most probable failure for the given structure (as example: Ansari, 2005; Douglas, 2007; Moustafa et al., 2010).

The methodological approach for the evaluation of a structure resistance is represented in Figure 1 where structural knowledge obtained through a series of test assessments is needed in order to define vulnerability and thus design suitable retrofit strategies.

*Romeo Bernini[2], Vincenzo Lapenna[1], Antonio Loperte[1], Francesco Soldovieri[2], Felice Carlo Ponzo[3], Antonio Di Cesare[3] and Rocco Ditommaso[3]
[1]CNR-IMAA, Italy
[2]CNR-IREA ,Italy
[3]Basilicata University/DiSGG, Italy

Since the level of reliability of the assessment method is related to the adequacy of the model and to the completeness of the information, all useful available data have to be collected in order to define the original structural characteristics such as: geometry of structural elements, characteristics and behaviour of the construction materials, presence of degradation, arrangement of longitudinal and transversal reinforcement.

The knowledge of an existing structure is never complete and the level and accuracy of construction details obviously corresponds proportionally with the available original design documentation, the time and funds available for in situ investigations and experimental tests on the structural elements.

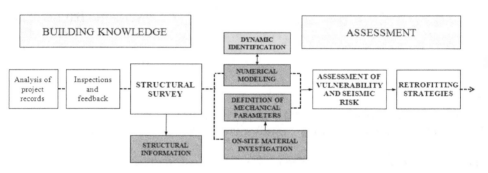

Fig. 1. Methodological approach.

A reliable assessment of the vulnerability of buildings is also strictly connected to the evaluation of the mechanical characteristics of the constitutive materials. This can be particularly complex for concrete, due to the high variability of its resistance that depends on intrinsic factors such as the composition as well as the environmental and maturing conditions, and other factors attributable to the collection technique and reworking of the concrete sample and the test conditions in general (Barlet, 1994).

A number of tests and methods have been developed for evaluating the resistance of construction materials ranging from completely non-destructive tests (NDT), where there is no damage to the structural element, using methods where the concrete surface is only slightly damaged, to partially destructive or destructive tests (DT), where the structural element has to be repaired afterwards.

The classical NDT methods, generally used for Reinforced concrete (R/C) structures, are the surface hardness method coupled with the ultrasonic method. As these methods are influenced in different and/or opposing ways by some fundamental parameters, their combined use allows outputs with minimal dispersion to be obtained. It is generally not advisable to use a single non-destructive test to estimate the strength in situ of concrete.

The range of properties that can be assessed using the range of NDT methods is significant and includes fundamental parameters such as density, elastic modulus as well as surface hardness, reinforcement location and depth of cover concrete. In some cases it is also possible to check the quality of workmanship and structural integrity through the ability to detect voids, cracking and delamination.

Preliminary tests can be performed with a covermeter according to the procedures described in the British Standard 188:204. With this technique it is possible to determine the presence and size of reinforcing bars, laps, transverse steel and depth and position of reinforcement.

The identification of the position of the reinforcement bars is also used as a preliminary to the other NDT (such as Ultrasonic Pulse Velocity, Schmidt Rebound Hammer, Pull-out UNI EN 12504-3, 2005) and also the DT (Core Extraction and Compression Test). In addition the partially destructive method of removing the cover concrete in some areas and measuring directly the diameter and type of the reinforcement can be performed.

The Schmidt rebound hammer test is principally a surface hardness tester and is carried out according to UNI EN 12504-2 (2001). The system works on the principle that the rebound of an elastic mass depends on the hardness of the surface against which the mass impinges. There is little apparent theoretical relationship between the strength of concrete and the rebound number of the hammer, however within limits, experimental correlations are established between strength properties and the rebound number. All of this cannot be generalized and should be calibrated for each type of existing concrete, for example using the results of compression tests. In the some cases the results of the hammer tests, taken as a rebound average (Ir) is used individually to assess the homogeneity of the concrete.

The Ultrasonic test is carried out in compliance with UNI EN 12504-4 (2005) and is aimed at determining the propagation velocity of a mechanical vibration pulse in concrete. By measuring the pulse crossing time and the distance between the two probes, the apparent propagation velocity can be calculated. This value can differ from the real value when the elastic waves undergo deviations from the path identified by the conjunction line between the two probes (RILEM 1972). The factors that affect the ultrasonic test the most are linked to the concrete composition, environmental and test conditions.

When interpreting the NDT results, special attention is needed regarding the presence of possible anomalies which can negatively influence the experimental assessment of the in situ concrete mechanical characteristics. Such anomalies are generally characterized through an evident correlation of the experimental datum with either a physical parameter of reference (usually compressive strength from DT) or with respect to the trend shown by the data acquired in the same context of structural homogeneity. These anomalies usually arise from improper execution of the test or from the fact that the test is carried out in non-ideal conditions.

Direct measure of the compressive strength of concrete in a structure is provided by the Concrete Core Extraction and Compression method (DT). The process of obtaining core specimens and interpreting the strength test results is often affected by various factors that influence either the in-place strength of the concrete or the measured strength of the test specimen (UNI EN 12504-1, 2002). In spite of such disturbance factors, values measured in this way are certainly the most reliable possible. Furthermore errors can be reduced using the A.C.I. 214.4R-03 guidelines which summarize current practices for obtaining cores and interpreting core compressive strength test results.

Immediately after extraction, the core concrete is tested for carbonation (also called depassivation). Carbonation penetrates below the exposed surface of concrete extremely slowly. The significance of carbonation is that the usual protection of the reinforcing steel generally present in the concrete due to the alkaline conditions caused by the hydrated cement paste is neutralized. Thus, if the entirety of the cover concrete is carbonated, corrosion of the steel will occur if moisture and oxygen can infiltrate the section.

The necessary destruction of the test object usually makes DT methods more expensive, and these testing methods can also be inappropriate in many circumstances. Therefore the use of NDT plays a crucial role in ensuring an economical operation. A general proportion of 1 core to 4 non-destructive investigations is recommended. Both the results from the DT and NDT are then combined in order to estimate the in situ concrete strength using the SonReb method. This method is the principal combination of Schmidt Rebound Hammer with Ultrasonic Pulse Velocity used for quality control and strength estimation of in situ concrete (Braga, 1992).

Another group of NDT methods are the Dynamic identification tests which can be used in order to assess fundamental dynamic properties (frequencies and/or modal shapes) of the structure and indirectly estimate the Young's modulus of the material (Ponzo et al., 2010). All dynamic characteristics can be estimated using two different approaches: classical methodologies based on Fourier analysis (Ditommaso et al., 2010a) or innovative methodologies based on time-frequency and interferometric analyses (Ditommaso et al., 2010b; Picozzi et al., 2011). These latter analyses are also useful to detect possible structural damage occurred after an earthquake (Ditommaso et al., 2011).

The results of the above testing methods (both NDT and DT) are used to calibrate numerical models. These models can then be compared to the likely seismic loading thus providing the overall vulnerability of the structure being considered (Ponzo et al. 2011).

Even if the methods described above can ensure (when correctly applied) a high level of structural knowledge a number of issues remain to be addressed. In order to do this innovative technologies and new methods must be developed: reduce uncertainty regarding core extraction points, improve the detection of deflection and deficiencies, detect water infiltration, improve reinforcement information, improve the depth under investigation, and reduce time and cost.

Due to rapid and flexible execution, high spatial resolution and deep investigation depth, electromagnetic sensing techniques are a group of NDT methods which can achieve these objectives. They can direct the use of classical NDT and DT methods and reduce uncertainties, coring number and the survey cost. Furthermore, their contribution to the structural knowledge allows the adoption of lower safety coefficients (through the increase in available information thus minimising spread) and thus higher calculation resistances. This in turn reduces the extent and cost of the actions required for the improvement or seismic retrofit of structures, if needed.

The building and infrastructure diagnostics can be take advantage from the use of new NDT techniques enabling larger investigation depths, spatial resolution, void and defect detection capacity, low cost and fastness. Electromagnetic sensing techniques can be an useful tool in order to achieve these objectives. They are based on injection of a form of electromagnetic energy (electrical current, radiowave, microwave, light, etc.) into the surveyed object and gathering of returned signal in order to measure electromagnetic properties (resistivity, electrical permittivity), reconstruct the inner structure, detect embedded defects.

In table 1 an overview of the advantages and disadvantages of several techniques (both classic and innovative) is presented, in terms of cost, speed of procedure, non intrusivity, accuracy of data obtained and degree of correlation with actual values.

Nevertheless, applying electromagnetic sensing techniques to man made structures, some adaptation needs to fit stringent requirements in terms of exploration depth, spatial resolution and signal/noise ratio. In fact, commonly used building materials pose challenging issues in terms of electrode impedance, coupling antennas, survey modalities, tomographic reconstruction, sensor size.

Electromagnetic sensing techniques suitable for civil infrastructures and building diagnostic such as Ground Penetrating Radar (GPR) and Electrical Resistivity Tomography (ERT) are presented in this chapter. Concerning the GPR we focus the attention on the possibility to improve the imaging at low and radio/microwave frequencies by using novel inversion approaches such as the Microwave Tomography (MT).
Then, a novel distributed fiber optic sensor technology able to monitor strain and temperature variations,   is described. Finally, we discuss about the real contribution provided by electromagnetic sensing techniques in the building and infrastructure monitoring.

Test Method	Cost	Speed of procedure	Damage effected	Accuracy of data	Degree of correlation
Core samples	High	Slow	Moderate	Moderate	High
Rebound hammer	Very low	Fast	None	Only surface info	Low
Ultrasonic waves	Low	Fast	None	Complete penetration	Moderate
Covermeter	Low	Fast	None	Moderate	-
Electromagnetic sensing	High	Moderate	None	Good	Low/ Moderate
Dynamic Test	Low	Moderate	None	Good	None/Low

Table 1. Advantages and disadvantages of different health assessment techniques.

2. Electromagnetic sensing techniques

Electromagnetic sensing techniques use is now rather diffuse in several earth science fields such as geology, hydrogeology, seismology, glaciology, stratigraphy of urban areas study, polluted areas study, landslides characterization, etc. Few years ago, non intrusiveness and quickness of these techniques suggested their use for investigating buildings and civil engineering structures.
The migration of these techniques towards the engineering can be identified with the Microgeophysics where specific issues are the sensor miniaturizing, signal/noise ratio improvement, exploitation of all available free surfaces for energizing and acquiring signals (Cosentino et al., 2011).
Electromagnetic sensing techniques are an useful tool for the diagnostics of civil infrastructures, such as transport ones, in the framework of their static and dynamic behavior before, during and after a crisis event such as an earthquake.
In fact, they ensure a fast and not-intrusive diagnosis useful in the pre-event stage since a precautionary diagnosis of strategic buildings and transport infrastructure can be a critical issue in the seismic risk prevention.
Moreover, they can represent, during the crisis, a valid tool for the rapid damage mapping of the civil buildings and infrastructures (bridges, roads, dams, assessment) in order to have preliminary estimations of those safe for rescue forces. Then, a rapid damage assessment for private buildings enable a correct estimate of the damaged houses and resources to be

committed. Finally, in the post-event stage, restoration interventions can be driven by the electromagnetic sensing techniques in order to minimize the costs and maximize the results.

The electromagnetic sensing techniques provides information about investigated materials in terms of amplitude and phase of the gathered signals, in turn function of the electromagnetic properties of the materials.

Amongst the requirements of the infrastructure diagnostics there is certainly the determination of the structural element thickness, rebar diameters, fractures and defects detection, water content or moisture (as indicators of chemical reactions occurrence), strain.

Therefore, the information obtained by the electromagnetic sensing techniques have to be converted in order to provide information directly usable by the engineers.

Not all electromagnetic sensing techniques are suitable in becoming NDT techniques. Sensing technique selection have to keep into account a certain degree of electromagnetic noise immunity, high spatial resolution and a suitable sensor size. Other aspects to be kept into account are:

- proper survey design;
- useful spatial resolution and investigation depth;
- suitable electrical contact, electromagnetic and/or mechanical coupling, precise positioning, boundary problems minimization;
- efficient data processing, including image processing and inversion techniques;
- easy data interpretation;
- possibility of data integration.

Ground Penetrating Radar (GPR) and Electrical Resistivity Tomography (ERT) have these qualities since they are active techniques providing the control on the injected signal, an adjustable spatial resolution and an useful sensor size. Moreover a number of processing codes and inversion routines are available allowing well interpretable images, although personnel with certain degree of experience is required in the data processing and interpretation.

Another class of sensors is the distributed ones, ensuring the availability of measurements along the entire envelop of the sensor. Among these, Fiber Optic Distributed Sensors based on Brillouin scattering phenomenon is a promising experimental technique able to provide field of temperature and strain along the fiber which can be a standard low cost telecommunication fiber. Unlike other fiber optic sensors, this technique permits the remote and spatially continuous monitoring of the structure in terms of temperature and strain with the resolution of some tens of centimetres.

In the following paragraphs we describe those three techniques providing examples of application and highlighting their strengths and limitations.

2.1 Ground penetrating radar

Ground Penetrating Radar is an electromagnetic sensing technique based on the same operating principles of classical radars (Daniels, 2004). In fact, it works by emitting an electromagnetic signal (generally modulated pulses or continuous harmonic waves) into the ground or another natural or manmade object; the electromagnetic wave propagates through the opaque medium and when it impinges on a non-homogeneity of the electromagnetic properties, in terms of dielectric permittivity and electrical conductivity, a backscattered electromagnetic field arises. Such a backscattered field is then collected by the receiving antenna located at the air/opaque medium interface and undergoes a subsequent processing and visualization, usually as a 2D image (Figure 2).

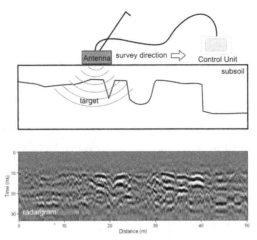

Fig. 2. GPR survey design (upper panel) and radargram (lower panel).

Spatial resolution and investigation depth of GPR method are strictly dependent by the central frequency of the used antenna. In fact, antennas with low and mid frequency (40 MHz - 750 MHz) provided high investigation depth (10 m - 3 m) associated to a relative low spatial resolution (2 m - 10 cm). On the contrary, high central frequency antennas (900 MHz - 2.5 GHz) provide low investigation depths (1 m - 10 cm) and high spatial resolution (5 cm - 0.5 cm). Since the physical size of the antennas decreases as the frequency increases, the requirement of miniaturizing the sensors is naturally achieved for the GPR.

As a consequence, the GPR technique is useful for the study of bedrock stratigraphy and cavity detection (Lazzari et al. 2006), groundwater and pollution (Chianese et al. 2006), metal and plastic pipelines such as cables in urban areas, archaeological finds (Bavusi et al. 2009) when low central frequency antennas are used.

On the contrary, when high central frequency antennas are used, the GPR, more properly named in this case Surface Penetrating Radar (SPR), can be considered a NDT technique (McCann and Forde 2001) providing precious information about the presence of "embedded" objects such as, reinforced rebars (Shaw et al., 2005; Che et al., 2009), but also embedded "defects" such as voids and, by using special antennas (Huston et al, 2000; Forest and Utsi, 2004; Utsi et al., 2008), fractures. Moreover, the GPR technique can contribute to determine the concrete moisture content (Shaari et al., 2004; Hugenschmidt and Loser, 2008;

GPR survey design is a crucial issue since it determines not only the possibility to detect the target (rebar, defect, water infiltration, ecc.), but also the format output in terms or 2D (cross section, time-slice, depth slice) or 3D data volume, kind of processing and difficult of interpretation.

A proper GPR survey design have to keep into account:

- orientation of searched target;
- possibility of exploit one or more suitable free surfaces;
- desired spatial resolution and investigation depth;
- mispositioning error minimization;
- format of the output;
- processing and interpretation effort minimization.

When rebars are searched, the most used method of acquisition requires a regular orthogonal survey grid with a proper spacing (a few centimeters) in order to have a suitable spatial resolution. Figure 3 shows a the survey design performed in order to check the continuity of longitudinal and transversal rebars and check the degree of success of the concrete restoration intervention based on epoxy resin injection (Bavusi et al., 2010a).

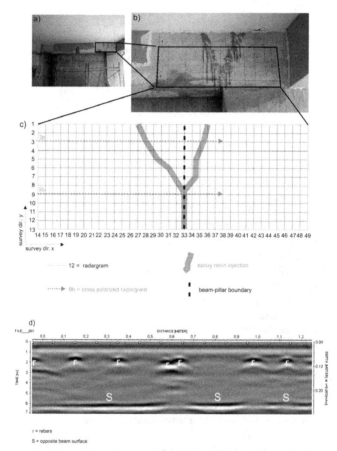

Fig. 3. GPR survey design and results on a beam of a school of L'Aquila damaged by the Abruzzo earthquake of 6th April 2009. a) beam; b) detail of damaged area with gridding; c) regular 4 cm square grid drawn on the beam; d) longitudinal processed radargram n.3.

A radargram has been gathered along each longitudinal and transversal survey line by using a 1500 MHz antenna provided by survey wheel. This survey design allows to select proper cross-sections and built a data volume. In fact, transversal radargrams offer a view of longitudinal rebars, while longitudinal radargrams are useful for visualizing transversal rebars. Then, the interpolation of all radargrams allows to built a data volume and selects more significant time-slices or depth-slices in order to have a plan view of all rebars (Figure 4).

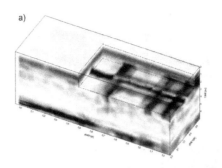

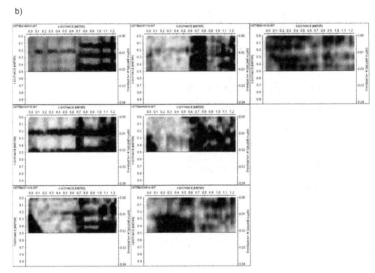

Fig. 4. a) data volume built by interpolating all radargrams gathered along the survey lines of figure 3c. b) slices extracted at several depths.

Then, GPR method provides very impressive and effective images of the inner of a reinforced concrete structure. However, the main limitation is that deeper rebar layer is not well detected due to scattering phenomena and attenuation losses producing in turn a loss in spatial definition in depth. Moreover the upper layer of rebars produces a strong disturbance on the rest of radargram.

In order to overcome this drawback, several strategies can be applied:

1. Repeating the measurements on the opposite surface of the structural element, if possible, in order to focus the other rebar layer;
2. Acquiring in cross-polarization mode: this method enables to focus the attention on the concrete matrix and not on the rebars;
3. Applying more robust inversion routines such as the Microwave Tomography.

First strategy can be effective, but increases the time consuming. The second one exploits the property of cross-polarized radargrams which are less sensitive to the rebars normal to the survey direction and more sensitive to the rebars parallel to the survey direction (Figure 5).

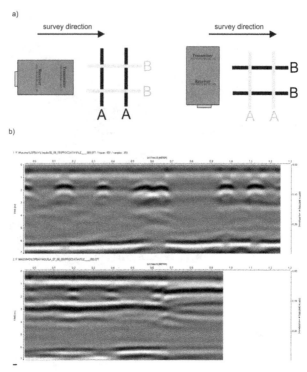

Fig. 5. a) Effect of the antenna polarization on the rebar reflection intensity. b) comparison between the normal-polarized and cross-polarized radargram gathered on the longitudinal survey lines n.9 of figure 2c.

Despite of the above mentioned advantages of GPR, one of the obstacles to its use regards the "low interpretability" of the radargram; therefore a an understandable "interpretation and visualization" of the investigated scene entails a high level operator's expertise and often a-priori information is required. This difficulty of the interpretation is further on affected in the case that no a priori information is available as, for example, it often happens in the case of historical heritage (Masini et al., 2010) where a lack of knowledge about the constructive modalities and materials of the structure can arise.

Therefore, a GPR data processing is often necessary to achieve more easily interpretable and reliable reconstructions of the scene, i.e. images that be easily understandable even by a not expert operator.

The usual radaristic approaches are based on migration procedures that essentially aim at reconstructing buried scattering objects from measurements collected above or just at the air/soil interface. These approaches were first based on graphical methods (Hagendoorn, 1954) based on high frequency assumption of the electromagnetic propagation and scattering; afterward this approach found a more consistent mathematical background based on the wave equation of the electromagnetic scattering (Stolt, 1978).

The absence of reflection in the concrete corresponding to the restored fracture indicates the success of the epoxy injection which filled all possible voids. Finally, a retrofit reinforcing intervention can be designed on the bases of the existent rebar arrangement.

Recently, new data processing based on the inverse scattering problem have been developed and implemented also in realistic situations for infrastructure monitoring (Catapano et al., 2006; Soldovieri and Orlando, 2009, Bavusi et al., 2011). In particular, the microwave tomography approaches have arose as the most suitable ones for the on field exploitation (Soldovieri and Solimene, 2010; Persico et al., 2005).

Such a class of approaches is based on the modeling of the electromagnetic scattering phenomena. According to this modelization, the imaging problem is cast as an inverse scattering problem where one attempts to infer the electromagnetic properties of the scattering object starting from the scattered field measured somewhere outside it.

The statement of the problem is then the following: given an incident field, E_{inc}, which is the electromagnetic field existing in the whole space (the background medium) in absence of the scattering object and is generated by a transmitting antenna, by the interaction of the incident field with the embedded objects the scattered field E_S arises; from the knowledge of the scattered field E_S properties about the scattering targets, either geometrical and/or structural, have to be retrieved. The mathematical equations subtending the scattering phenomena to solve the above stated problem are in order.

To this end, we refer to a two-dimensional and scalar geometry. We consider a cylindrical dielectric object (i.e. invariant along the axis out-coming from the sheet) enclosed within the domain D illuminated by an incident field linearly polarized along the axis of invariance. The scattered field is observed over the domain Σ (not necessarily rectilinear). Moreover, we denote by $\varepsilon(\underline{r})e$ by $\varepsilon_b(\underline{r})$ the permittivity profile of the unknown object and of the background medium, respectively. In particular, the latter is not necessarily constant (i.e., a non-homogeneous background medium is allowed too) but has to be known. The magnetic permeability is assumed equal to that of the free space μ_0 everywhere. The geometry of the problem is detailed in Figure 6.

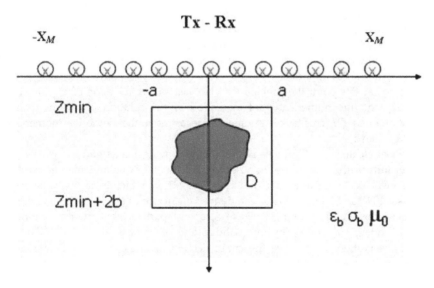

Fig. 6. Geometry of the subsurface prospecting problem

The problem, thus, amounts to retrieving the dielectric permittivity profile $\varepsilon(\underline{r})$ of the unknown object(s) from the knowledge of the scattered field E_S. The physical phenomenon is governed by the two equations (Chew, 1995)

$$E(\underline{r},\underline{r}_S;k_b) = E_{inc}(\underline{r},\underline{r}_S;k_b) + k_b^2 \int_D G(\underline{r},\underline{r}';k_b)E(\underline{r}',\underline{r}_S;k_b)\chi(\underline{r}')d\underline{r}' \qquad \underline{r} \in D$$

$$E_S(\underline{r}_O,\underline{r}_S;k_b) = k_b^2 \int_D G(\underline{r}_O,\underline{r};k_b)E(\underline{r},\underline{r}_S;k_b)\chi(\underline{r})d\underline{r} \qquad \underline{r}_O \in \Sigma \tag{1}$$

where $E = E_{inc} + E_S$ is the total field, k_b is the subsurface (background) wave-number and $\chi(\underline{r}) = \varepsilon(\underline{r}) / \varepsilon_b - 1$ is the dimensionless contrast function. $G(\bullet,\bullet)$ is the pertinent Green's function (Leone and Soldovieri, 2003), $\underline{r}_O$ is the observation point and $\underline{r}_S$ is the position of the source.

In accordance to the volumetric equivalence theorem (Harrington, 1961), the above integral formulation permits to interpret the scattered field as being radiated by secondary sources (the "polarization currents") which are just located within the space occupied by the targets. The reconstruction problem thus consists of inverting the "system of equations (1)" versus the contrast function. However, since (from the first of the equations 1) the field inside the buried targets depends on the unknown contrast function, the relationship between the contrast function and the scattered field is nonlinear. However, the problem can be cast as a linear one if the first line equation is arrested at the first term of its Neumann expansion. After doing this $E \cong E_{inc}$ is assumed within the targets and the so-called Born linear model is obtained (Chew, 1995). Accordingly, the scattering model becomes

$$E_S(\underline{r}_O,\underline{r}_S;k_b) = k_b^2 \int_D G(\underline{r}_O,\underline{r};k_b)E_{inc}(\underline{r},\underline{r}_S;k_b)\chi(\underline{r})d\underline{r} \qquad \underline{r}_O \in \Sigma \tag{2}$$

Let us just remark that, within the linear approximation, the internal field does not depend on the dielectric profile, which is the same as to say that mutual interactions between different parts of any object or between different objects are neglected. In other words, this means to consider each part of the object as an elementary scatterer that does not depend on the presence of the other scatterers.

Consequently, at this point the problem can be cast as the inversion of the linear integral equation (2) and the numerical implementation of the solution algorithm requires the discretization of eq. (2). This task is pursued by resorting to the method of moments (MoM) (Harrington, 1961).

One of the main feature of a GPR data is its ability to provide images of the inner structure of a building or infrastructure at all useful observation scales by exploiting antennas with several central frequencies. Concerning this, the bridge inspection, which is normally carried out by using classical DT and NDT methods, can derive benefit from the GPR technique (Scott et al., 2003; Hugenschmidt and Mastrangelo, 2006). Structural particulars of interest are inner rebars, tendons, boundary conditions, anchors, saddles and other internal elements. On the other hand, the observation can involve the entire deck of a bridge. Such observations became crucial when all original project documentations are partially or completely lost.

Figure 7 shows the survey design drown for the deck survey of the deck of Musmeci bridge in Potenza (Basilicata Region, Southern Italy) (Bavusi et al., 2011).

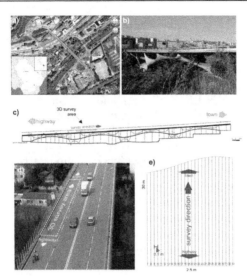

Fig. 7. a) location of the Musmeci bridge in Potenza, Basilicata Region, Italy. b) location of the survey area respect to the bridge; d) surveyed lane; e) survey grid. From Bavusi M., Soldovieri F, Di Napoli R., Loperte A., Di Cesare A., Ponzo F.C. and Lapenna V (2010).Ground Penetrating Radar and Microwave Tomography 3D applications for the deck evaluation of the Musmeci bridge (Potenza, Italy). J. Geophys. Eng. 8 (2011) 1–14. Courtesy of IOP Publishing Ltd.

A such survey design is able to provide several depth-slices in order to observe the asphalt layer, the sects layer and the lower reinforced concrete plate of the deck. Figure 8 shows more significant depth-slices showing the deck structure at several depths.

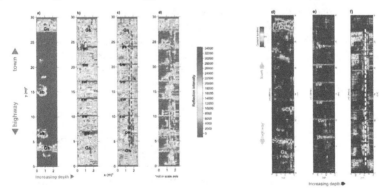

Fig. 8. a-d) depth-slices built at 6 cm (a), 20 cm (b), 40 cm (c) and 50 cm (d) by interpolating all radargrams gathered on the survey lines of figure 5e; d-f) depth-slices built at 6 cm (d), 20 cm (e) and 40 cm (f), by interpolating the same radargrams inverted by means of the Microwave Tomography. ab: absorption zone, Gs: Gerber saddle, ps: pillar support, sw: stiffening wall, ls: longitudinal stiffening wall. From Bavusi M., Soldovieri F, Di Napoli R., Loperte A., Di Cesare A., Ponzo F.C. and Lapenna V (2010).Ground Penetrating Radar and Microwave Tomography 3D applications for the deck evaluation of the Musmeci bridge (Potenza, Italy). J. Geophys. Eng. 8 (2011) 1–14. Courtesy of IOP Publishing Ltd.

In particular the asphalt layer shows concentrated absorptive zones that can be related to water infiltration zones. In this case the precise positioning of traces is a critical issue since it can produce a staggering effect when rectilinear features are detected. The use of a survey wheel is mandatory such as a certain degree of care in the cart dragging in order to limit mispositioning errors. In this way a residual error can be subsequently reduced by using proper algorithms.

Among possible defects afflicting buildings and infrastructures, fractures are a very warring problem. In fact, fractures can be due to several causes: temperature (3 mm), dry up (0.4 mm), load (0.4 mm). They can involve loss of mechanical strength and represent a preferential way for the water infiltration which in turn can favour the developin g of chemical reactions (expanding salt crystallization, oxidation, carbonation, etc.).

Crack detection is an important issue in the field of non-destructive testing. Several techniques can be employed in order to check, localize and characterize fractures in manmade buildings: ultrasonic shear waves (De La Haza et al. 2008), elastic waves (Ohtsu et al. 2008), GPR (Utsi et al. 2008).

Due to their small size and variable orientation, fractures detection represents a very challenge for the GPR technique. In fact, the crack detection requires the exploitation of all spatial resolution available in the frequency range used. Moreover it requires fracture to be surveyed is filled by air, water or a material different from the host medium in order to produce a backscattered field (Grandjean and Gourry 1996). In addition, the geometry of the fracture with respect to the survey line plays a fundamental role (Tsoflias et al. 2004). For a vertically oriented fracture, a reflection hyperbola arises due to the bottom of the fracture and to each change in the direction of the fracture with respect to the vertical path (Forest and Utsi 2004). By exploiting this property, it is theoretically possible to detect a fracture by using a common GPR dipole antenna, even the use of specifically designed high vertical resolution antennas is very helpful (Forest and Utsi 2004; Utsi et al. 2008). Moreover, data processing plays a fundamental role to improve the 'imaging' and the focusing of the buried reflectors (Grandjean and Gourry 1996; Leucci et al. 2007).

Figure 9 shows a 1500 MHz GPR survey carried out on a fracture in the floor of the Prefecture of Chania (Crete Island, Greece) (Bavusi et al., 2010b).

Fracture zone, located at the middle point of the radargram, is detected by using a classical processing approach, but best performances in terms of spatial resolution can be obtained by using the Microwave Tomography.

2.2 Electrical resistivity tomography

Electrical resistivity tomography (ERT) is an electromagnetic sensing technique used to obtain 2D and 3D images in terms of electrical resistivity of areas of complex geology (Griffiths and Baker 1993), landslides, watertable, basins, faults.

Technically, during an electrical resistivity measurement, the electric current is injected into the ground via two 30-40 cm × 1.5 cm steel electrodes and the resulting electrical voltage is measured between two other electrodes in line with current electrodes (Sharma, 1997). ERT can be carried out by using different electrode configurations such as dipole-dipole, Wenner, Schulumberger, pole-dipole, etc. (Figure 10).

At present, such configurations can be carried out by using multi-electrode systems enabling the automatic switch of all available electrodes previously fixed into the ground. The system manages the current injection and simultaneous potential measurements which can occur simultaneously at more potential electrodes in case of multichannel systems.

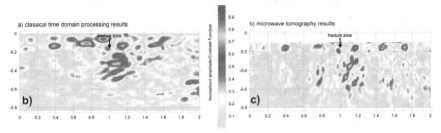

Fig. 9. a) GPR survey design carried out at the Prefecture of Chania (Crete, Greece) on a fracture in the floor. Fracture zone is at the middle point of the radargrams. b) processed radargram; c) Microwave Tomography From Bavusi M., Soldovieri F., Piscitelli S., Loperte A., Vallianatos F. and Soupios P. (2010). Ground-penetrating radar and microwave tomography to evaluate the crack and joint geometry in historical buildings: some examples from Chania, Crete, Greece. Near Surface Geophysics, Vol.8, No. 5, pp. 377-387. Courtesy of EAGE Publications BV..

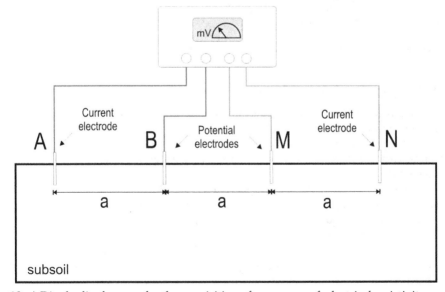

Fig. 10. a) Dipole-dipole array for the acquisition of a measure of electrical resistivity.

The result of an ERT survey is a distribution 2D or 3D of apparent resistivity where each data point is defined by two coordinates (x and z) depending on the position of the quadrupole (couple of current and potential dipoles) used and a value of apparent resistivity. Then, in order to reconstruct real resistivity distribution, an inversion routine is required. A number of algorithms are available in order to perform this reconstruction such as Res2DInv (Loke and Barker, 1996) for the automatic 2D inversion of apparent resistivity data was used. The inversion routine is based on the smoothness constrained least-squares inversion (Sasaki, 1992) implemented by a quasi-Newton optimization technique.

ERT surveys have been successfully applied in geology for stratigraphy and cavity detection (Lazzari et al., 2010), fault characterization (Caputo et al., 2007), landslide studies (Lapenna et al., 2005), in hydrogeology, in environmental problems for contaminant plume detection and waste dump characterization (Bavusi et al., 2006), for hydrogeology and coastal salt water intrusion detection (Satriani et al., 2011a), in agricultural for the root-zone characterization (Al Hagrey, 2007; Satriani et al, 2011b), in archaeology and cultural heritage studies (Bavusi et al., 2009). Figure 11 shows an example of ERT carried out on a piling in an area subjected to landslides. It is well visible the effect of the structure on the water distribution.

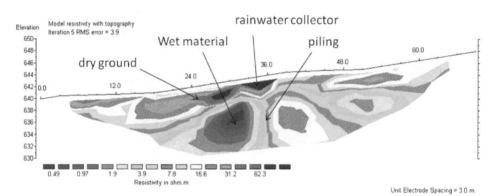

Fig. 11. Example of ERT carried out on a piling in an area subjected to landslides.

The ERT exhibits significant potentialities in terms of high resolution and flexibility of the investigation depth that can be varied in a simple way, by varying the electrode spacing. A large electrode spacing provides a high investigation dept and a low spatial resolution. On the contrary, a small electrode spacing allows to achieve a great spatial resolution but a low investigation depth.

This characteristic makes the ERT a candidate for the structure and infrastructure monitoring, even though some problems have to be enfaced. First, a structure or infrastructure survey requires an electrode spacing ranging between one centimeter and some decimeter, then sensors have to be miniaturized in order to respect the assumption of geophysics stating transducers have to be punctual (i.e. small respect to dimension of the investigated volume) and reduce modeling errors (Cosentino et al. 2011; Athanasiou et al., 2007). Then, a low contact resistance have to be ensured in order to put an adequate current injection (Cosentino et al. 2011).

In order to meet these requirements, several devices can be used such as Cu flat-base electrodes with conductive gel (Athanasiou et al., 2007), Ag/AgCl medical electrodes and nails (Cosentino et al 2011), Cu-CuSO electrodes (Seppänen et al., 2009). Main limitation of these devices is the difficulty of put them on a vertical or steeply slope surface, even worse under a ceiling. Moreover, medical electrodes are not stable in time (Cosentino et al 2011), while flat-base electrodes are not suitable for the asphalt, where the only possibility to apply the ERT is making holes in order to put the electrode in the substratum.

In spite of these limitations, the ERT has been successfully applied on masonry, floors, artifacts in order to detect fractures, voids, previous restoration works, structural particulars, moisture. The application to reinforced concrete is possible in order to detect those targets and rebars, but experiments demonstrates target detection capacity in simple geometrical configurations (Seppänen et al., 2009; Karhunen et al., 2009).

In presence of complex rebar configurations such as reinforcement cages, the potential field produced by the injected current suffers a warping due to the circuit represented by the cage not easily modelable. A new class of inversion routines is then required in order to solve this problem. This technique appears still not adequate to be applied to structures and infrastructures, but technological development can provide technical solutions able to mitigate and overcame described limitations.

2.3 Distributed fiber optic sensors

Typically standard NDT systems are based on the use of point sensors, however the Structural Health Monitoring (SHM) of large civil infrastructures like bridges, dams could require a large number of sensors. For these applications, there is an increasing interest towards the use of distributed optical fiber sensors. In these sensors is the optical fiber itself that acts as a sensor providing measurements all along the fiber. This approach permits to monitoring the whole structure by use of a single optical fiber avoiding the need of a huge amount of measurement points and lead to the comprehension of the real static behavior of the structure rather than a limited number of sensors. Furthermore, distributed sensors could play a fundamental role in civil engineering because no other tools allow the detection of local phenomena whose location is impossible to be predicted "a priori" like, for instance, for crack detection. Distributed fiber optic sensors are substantially different from other fiber optic sensors technologies being based on optical scattering mechanism (Raleigh, Raman, Brillouin) occurring during light propagation in common telecom optical fibres.

Spatial resolution is typically achieved by using the optical time domain reflectometry (OTDR) (Barnoski 1976), in which optical pulses are launched into an optical fiber and consequent variations in backscattering intensity is detected as a function of time. Alternative detection techniques such as frequency-domain approaches have been also demonstrated. Raleigh scattering based sensors were first developed, in order to locate fiber breaks or bad splices along a fiber link. However Rayleigh backscatter in standard fibers gives information only about optical attenuation, and it can not be related to other parameters such as temperature or strain. Distributed temperature sensing was first demonstrated by Hartog and Payne (1982), who used temperature-induced variation of the Rayleigh scattering coefficient along the length of liquid-core, but the low reliability of liquid-core fibers may restrict their use. A recent, very interesting approach makes use of the very high spatial resolution allowed by swept-wavelength interferometry, in order to correlate temperature and strain of the fiber with the spectrum of the Rayleigh backscatter spatial fluctuations (Measures 2001). This approach requires standard telecommunication fibers and very high spatial resolution (a few millimeters) has been demonstrated. On the other hand, the main disadvantages are the equipment cost (a tunable laser source is needed for the measurements), and the limited sensing length (70m). Dakin et al. (1985) demonstrated temperature profiles measurement using the variations in the Raman backscattering coefficients of anti-Stokes and Stokes light. The Raman approaches are very practical because conventional silica-based optical fibers can be used as the sensor. The anti-Stokes-Raman-backscattered light is about 30 dB weaker than the Rayleigh-backscattered light. However, its sensitivity to temperature is great. Therefore, systems based on Raman scattering have been commercialized by several manufacturers. Nevertheless, Raman scattering based systems do not allow performing deformation measurements. In 1989, it was reported that the frequency shift of the Stokes-Brillouin-backscattered light (the so-called Brillouin frequency shift) greatly varies with strain and temperature along the fiber (Horiguchi 1989, Culverhouse 1989). Since then, considerable attention has been paid to exploiting Brillouin scattering for distributed sensing. This is for the following reasons. First, strain is a very important parameter in the monitoring of the integrity of civil structures. Secondly, unlike the Raman technique, Brillouin frequency shift measurement does not require calibration of the optical-fiber loss. Furthermore, a very attractive feature of Brillouin-based sensors stems from the use of a standard telecommunications-grade optical fiber as the sensor head. The low-cost and low-loss nature of the sensor make possible to perform distributed measurements over distances of many kilometers. Finally, the tremendous developments in the optical telecommunications market have reduced considerably the cost and increased the performances of optical fibers and their associated optical components.

Distributed optical fiber sensors based on stimulated Brillouin scattering (SBS) rely on the interaction between two lightwaves and an acoustic wave in the optical fiber. The measurement principle is based on the characteristic that the Brillouin frequency of the optical fiber is shifted when strain as well as temperature changes occur. Spatial information along the length of the fiber can be obtained through Brillouin optical time domain analysis (BOTDA) by measuring propagation times for light pulses travelling in the fiber. This allows for continuous distributions of the parameter to be monitored.

The research in the beginning of distributed fiber optic strain sensing was mostly based on laboratory applications (Bernini 2005, Bernini 2008) only in last years in-field demonstration by a fully distributed sensor have been previously reported (Komatsu 2002).

About bridge structures recently the applications and the validation of distributed strain sensor during load test has been demonstrated (Matta 2008, Minardo 2011). Other examples are the installation of a distributed fiber optic strain sensing cable into the inspection gallery of a dam (Inaudi and Glisic 2005, Glisic and Inaudi, 2007) or the monitoring of extra-long tunnel, running over 150 km of seafloor geologic body with complicated topographic and geologic units (Shi, 2003).

However, the use of distributed fiber optic sensors for crack detection in concrete are rare. This is mainly due to the fact that the instrumentations available for in-field application have a limited spatial resolution (1m) (Deif 2010). In fact, distributed sensors measure the with average strain at each measurement point, where the strain is averaged over the length called spatial resolution. Today, new methods for distributed fiber optic strain sensing with sub-meter spatial resolution are being developed in order to increase the opportunities in NDT of civil structures especially for crack detection (Hotate 2002, Zou 2005, Bernini 2007).

As an example, we report the results obtained in a load test on a road-bridge (Minardo 2011). In particular, the tests were performed by an stimulate Brilouin Scattering portable sensor prototype with 3m-spatial resolution. The fiber employed for the measurements was a PVC-coated, single-mode, standard telecom optical fiber. The fiber was bonded along the lower flange of a 44-m-long, double-T steel beam, by use of a epoxy adhesive. Strain measurements were performed while loading the bridge with an increasing weight by use of gravel-loaded trucks (Figure 12a). During the loading test, data were also collected by other instruments for a cross-correlation. In particular, two vibrating wire (VW) strain gauges were previously spot-welded to the surface of the steel beam, so as to provide the strain at the quarter and the middle section of the loaded beam. Figure 10b depicts the results of the optical fiber measurements, for different load conditions. In particular, the solid lines refer to the bridge loaded by two, four, and five gravel trucks, respectively. Each truck had a weight of approximately 47 tons. The same figure also reports the results of a finite-element-method (FEM)-based numerical analysis (circles). Numerical data were obtained by modelling each gravel truck as three concentrated loads applied in correspondence of the three truck axles. A good agreement exists between the experimental and numerical data. The standard deviation was always less than 20 µε, corresponding to the nominal accuracy of the instrument. Moreover, the maximum strain provided by the optical fiber sensor ($\approx$ 350 µε), is in good agreement with the value provided by the strain gauge placed at the middle beam section ($\approx$330 µε). Another interesting feature is that the optical fiber sensor was able to reveal the right-shift of the center of gravity (CG), when loading the beam with five trucks. Actually, the fifth truck was not disposed symmetrically with respect to the middle of the bridge; rather it was closer to the right side. As a consequence, the section at which maximum strain occurs is shifted to the right in the final load test. Finally, in figure 12b are also reported, for comparison, the data obtained by the strain gauges (squares). As can be observed a good agreement between the two different observations is achieved.

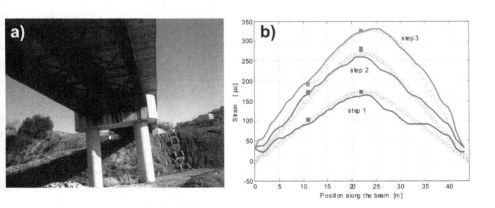

Fig. 12. a) bridge used for the load test. b) Distributed strain measurement along the girder, for different load conditions (solid lines). FEM simulations (open circles). VW strain gauges (squares).

3. Conclusion

Health monitoring of civil engineering structures is devoted to assess the structural integrity and dynamic behaviour during a seismic event in order to define appropriate activities for risk mitigation. This need is more stringent for aged infrastructures built following outdated codes of practice and where the chemical degradation of concrete and irons acted for more time.

The design of suitable retrofit strategies requires a series of test assessments in order to define the vulnerability. A number of destructive (DT) and non destructive tests (NTD) and methods has been developed and applied around the world in order to achieve all necessary information about construction material. An appropriate survey have to include a suitable proportion of DT and NDT methods in order to reduce possible damage to investigated structure, achieve a balanced combination of punctual and distributed information and reduce the global cost of the survey.

Then, the requirement of new non invasive technique is a stringent need that can be satisfied by the electromagnetic sensing techniques, this class of geophysical techniques which can be easily adapted to the specific requirements of civil infrastructures. This is the case of the Ground Penetrating Radar which at high frequencies provides the needed spatial resolution and a convenient small size of the antenna. Moreover, this technique can benefit on a new kind of processing based on the Microwave Tomography (MT) inversion. In this way the technique can focus small defects such as fractures and voids, detect rebars with great precision even if deep. The combination of GPR and MT will provide in future a new class of devices able to supply a not focused image ready for use by non expert users.

The Electrical Resistivity Tomography can be another suitable technique useful to depict embedded structural particulars and defects, but its systematic application requires to solve two problems. First is the design of appropriate non intrusive electrodes, stable in time and easy to install in any position. A solution can be provided by medical industries that have experience in the design of electrodes for human body applications. The latter is the lack of inversion routines able to model the effect of a rebar cage on the potential field and properly reconstruct the inner of a reinforced concrete structure. Anyway the ERT is successfully applied on floors, pavements, masonry and reinforced concrete structures having a simple inner arrangement. Moreover, the industry will provide in future new sensors, equipments and inversion routines able to mitigate or solve described problems.

Another class of sensors which will change the way of monitor a civil infrastructure is the distributed fiber optic sensor able to provide temperature and strain information along the fiber. This sensor is based on standard telecom fiber optic which is inexpensive and allows to design sensor sized on the infrastructure to be surveyed or on a particular. It allows to achieve in real time information by several points of a distributed civil infrastructure such as a railway, aqueduct, gas or oil pipeline without using transmission devices. In future optic fibers will be embedded in construction materials allowing a for life monitoring of an infrastructure. Moreover, a possible technological improvement of this technique can allow in future its application to the ambient vibration monitoring (Wenzel and Pichler, 2005).

4. Acknowledgment

The research leading to these results concerning the Musmeci bridge has received funding from the European Community's Seventh Framework Programme (FP7/2007-2013) under Grant Agreement n° 225663 Joint Call FP7-ICT-SEC-2007-1.
Moreover, the authors would like to thank the "Soprintendenza per i Beni Architettonici e Paesaggistici della Basilicata" and the "Direzione Regionale per i beni Culturali e Paesaggistici della Basilicata" that have partially funded the activities. Finally, the authors would like to thank the owner of the Musmeci Bridge, ASI Consortium, for their interest to these research activities and the Municipality of Potenza for the granted authorization to access and work on the structure. Furthermore, the authors would like to thank EAGE Publications BV and IOP Publishing Ltd for copyright permissions granted.
Finally, the authors would like to thank TeRN Consortium for supporting this work.

5. References

A.C.I. 214.4R (2003). Guide for Obtaining Cores and Interpreting Compressive Strength Results.

Al Hagrey, S.A., (2007). Geophysical imaging of root-zone, trunk, and moisture heterogeneity, *J. Exp. Bot.*, Vol. 58, pp. 839–854.

Ansari F.(2005). Sensing issues in civil structural health monitoring. Spinger.

Athanasiou E.N. , Tsourlos P.I., Vargemezis G.N., Papazachos C.B. and

Barlet, F.M. & MacGregor, J. (1994). Effect of Core Diameter on Concrete Core Strengths. *ACI Materials Journal*, Vol. 91, No. 5, September-October.

Barnoski J. K., Jensen S. M., (1976). Fiber waveguides: A novel technique for investigation attenuation characteristics. *Appl. Opt.*, Vol. 15, pp. 2112 - 2115.

Bavusi M., Rizzo E., Lapenna V. (2006). Electromagnetic methods to characterize the Savoia di Lucania waste dump (Southern Italy). *Environmental Geology*, Vol. 51, No. 2, pp. 301-308. DOI 10.1007/s00254-006-0327-9

Bavusi M., Rizzo E., Giocoli A., Lapenna V., (2009). Geophysical characterisation of Carlo's V Castle (Crotone, Italy). *Journal of Applied Geophysics*, Vol. 67, No. 4, ,pp. 386-401.

Bavusi, M., Loperte, A.; Lapenna, V., Soldovieri, F., (2010a). Rebars and defects detection by a GPR survey at a L'Aquila school damaged by the earthquake of April 2009., 2010 *13th International Conference on Ground Penetrating Radar (GPR),* 21-25 June 2010.

Bavusi M., Soldovieri F., Piscitelli S., Loperte A., Vallianatos F.and Soupios P. (2010b). Ground-penetrating radar and microwave tomography to evaluate the crack and joint geometry in historical buildings: some examples from Chania, Crete, Greece. *Near Surface Geophysics*, Vol.8, No. 5, pp. 377-387.

Bavusi M., Soldovieri F., Di Napoli R., Loperte A., Di Cesare A., Ponzo F. C. and Lapenna V. (2011). Ground Penetrating Radar and Microwave Tomography 3D applications for the deck evaluation of the Musmeci bridge (Potenza, Italy). *Journal of Goph. and Eng,* No. 8, pp. 1–14.

Bernini R., Fraldi M., Minardo A., Minutolo V., Carannante F., Nunziante L, Zeni L. (2005). Damage detection in bending beams through Brillouin distributed optic-fibre sensor. *Bridge Struct.* Vol. 1, 355–363.

Bernini R., Minardo A., Zeni L., (2007). Accurate high-resolution fiber-optic distributed strain measurements for structural health monitoring. *Sensors and Actuators A*, Vol. 134, pp. 389-395.

Bernini R., Minardo A., Zeni L. (2008). Vectorial dislocation monitoring of pipelines by use of Brillouin-based fiber-optics sensors. *Smart Mater. Struct.*, Vol. 17, pp. 015006,

Braga, F., Dolce, M., Masi, A., Nigro, D. (1992). Valutazione delle caratteristiche meccaniche dei calcestruzzi di bassa resistenza mediante prove non distruttive. *L'Industria Italiana del Cemento*, Vol. 3, pp. 201-208.

Caputo R., Salviulo L., Piscitelli S., Loperte A., (2007). Late Quaternary activity along the Scorciabuoi Fault (Southern Italy) as inferred from electrical resistivity tomographies. Annals of Geophysics, Vol. 50, No., 2, pp.213-224.

Catapano I., Crocco L., Persico R., Pieraccini M., Soldovieri F. (2006). Linear and Nonlinear Microwave Tomography Approaches for Subsurface Prospecting: Validation on Real Data. *Antennas and Wireless Propagation Letters*, Vol. 5, pp. 49-53.

Che W. C., Chen H. L. and Hung S. L. (2009). Measurement radius of reinforcing steel bar in concrete using digital image GPR. *Constr. Build. Mater.*, Vol. 23, pp. 1057–1063.

Chew C. W.(1995). *Waves and Fields in inhomogeneous media.* Piscataway,NJ: IEEE Press.

Chianese D., D'Emilio M., Bavusi M., Lapenna V., Macchiato M. (2006). Magnetic and ground probing radar measurements for soil pollution mappingin the industrial area of Val Basento (Basilicata Region, Southern Italy): a case study. *Environ Geol* , Vol. 49, pp. 389–404.

Cosentino P.L., , Capizzi P., Martorana R., Messina P. and Schiavone S. (2011). From Geophysics to Microgeophysics for Engineering and Cultural Heritage. International journal of Geophysics, Vol. 2011, pp. 1-8. DOI: 10.1155/2011/428412.

Culverhouse D., Fahari F., Pannell C. N., Jackson D. A.(1989). Potential of stimulated Brillouin scattering as sensing mechanism for distributed temperature sensors. *Electron. Lett.*, Vol. 25, No. 14, pp. 913-915.

Dakin J. P., Pratt D.J, Bibby G. W., Ross J. N., (1985), Distributed optical fiber Raman temperature sensor using a semiconductor light source and detector, *Electron. Lett.*, Vol. 21, No. 13, pp. 569-570.

Daniels D.J. (2006). *Ground Penetrating Radar.* 2nd edition. Institution of Engineering and Technology, pp.634. DOI: 978-0-86341-360-5; electronic DOI: 978-1-59124-893-4

De La Haza A.O., Petersen C. G. and Samokrutov A. (2008). Three dimesional imaging of concrete structures using ultrasonic shear waves. *Structural Faults & Repair-2008, 12th International Conference*, Edinburgh, UK: 10th -12th June 2008.

Deif A., Martın-Perez B., Cousin B., Zhang C., Bao X., Li W. (2010). Detection of cracks in a reinforced concrete beam using distributed Brillouin fibre sensors. *Smart Mater. Struct.*, Vol. 19, 055014.

Ditommaso R., Parolai S., Mucciarelli M., Eggert S., Sobiesiak M. and Zschau J. (2010a). Monitoring the response and the back-radiated energy of a building subjected to ambient vibration and impulsive action: the Falkenhof Tower (Potsdam, Germany). *Bulletin of Earthquake Engineering.* Vol..8, No. 3. DOI: 10.1007/s10518-009-9151-4.

Ditommaso R., Mucciarelli M., Ponzo F. C. (2010b). S-Transform based filter applied to the analysis of non-linear dynamic behaviour of soil and buildings. *14th European Conference on Earthquake Engineering.* Ohrid, Republic of Macedonia, August 30 – September 03.

Ditommaso R, Mucciarelli M., Ponzo F. C. (2011). Analysis of non-stationary structural systems by using a band-variable filter. Submitted to Bulletin of Earthquake Engineering.

Douglas A. (2007). Health Monitoring of Structural Materials and Components: Methods with Applications. Copyright © 2007 John Wiley & Sons, Ltd.

Forest R. and Utsi V. (2004). Non destructive crack depth measurements with ground penetrating radar. *10th Int. Conf. on Ground Penetrating Radar,*Delft, The Netherlands, 21–24 June 2004.

Glisic B. and Inaudi D. (2007). Fibre Optic Methods for Structural Health Monitoring. John Wiley & Sons. ISBN: 0470061421.

Grandjean C. and Gourry J.C. (1996). GPR data processing for 3D fracture mapping in a marble quarry (Thassos, Greece). *Journal of Applied Geophysics,*Vol. 36, pp.19-30.

Griffiths, D.H. and R.D. Baker (1993): Two-dimensional resistivity imaging and modelling in areas of complex geology. *J. Appl. Geophys.*, Vol. 29, pp. 211-226.

Hagedoorn, J. G. (1954). A Process of Seismic Reflection Interpretation. *Geophys. Prospect.* Vol. 2, pp 85-127.

Harrington, R.F. (1961). *Time-Harmonic Electromagnetic Fields*, Mc Graw Hill.

Hartog A. H., and D. N. Payne (1982). Remote measurement of temperature distribution using an optical fiber. *Proc. ECOC '82*, Cannes, France, pp. 215-220, 1982.

Horiguchi T, Kurashima T., Tateda M., (1989). Tensile strain dependence of Brillouin frequency shift in silica optical fibers. *IEEE Photonics Tech. Lett.*, Vol. 1, No. 5, pp. 107-108.

Hotate K., Tanaka M., (2002). Distributed fiber Brillouin strain sensing with 1-cm spatial resolution by correlation-based continuous-wave technique. *IEEE Phot. Technol. Let.* Vol. 14, 179-181.

Hugenschmidt J. and Mastrangelo R. (2006). GPR inspection of concrete bridges *Cem. Concr. Compos.*, Vol. 28, pp. 384–92.

Hugenschmidt J. and Loser R. (2008). Detection of chlorides and moisture in concrete structures with ground penetrating radar. *Mater. Struct.* Vol. 41, pp. 785–92

Huston D., Hu J. Q., Maser. K., Weedon W. and Adam C. (2000). GIMAground penetrating radar system for monitoring concrete bridge decks. *J. Appl. Geophys.*, Vol. 43, pp. 139–46.

Inaudi D., Glisic B., (2005). Application of distributed Fiber Optic Sensory for SHM", *2nd International Conference on Structural Health Monitoring of Intelligent Infrastructure* (SHMII-2'2005), Shenzhen, China, November 16-18.

Karhunen K., Seppänen A., Lehikoinen A.&. Kaipio J.P,(2009). Locating reinforcing bars in concrete with Electrical Resistance Tomography. *Concrete Repair, Rehabilitation and Retrofitting II - Alexander et al (eds).* © 2009 Taylor & Francis Group, London, ISBN 978-0-415-46850-3

Komatsu, K., Fujihashi, K. & Okutsu, M., (2002). Application of optical sensing technology to the civil engineering field with optical fiber strain measurement device (BOTDR). *Proc. SPIE,* Vol. 4920: 352-361.

Lapenna V., Lorenzo P., Perrone A., Piscitelli P., Rizzo E. Sdao F., (2005). 2D electrical resistivity imaging of some complex landslides in the Lucanian Apennine chain, southern Italy. *Geophysics,* Vol. 70, No. 3; pp. B11-B18. DOI: 10.1190/1.1926571

Lazzari M., Loperte A., and Perrone A. (2006). Near surface geophysics techniques and geomorphological approach to reconstruct the hazard cave map in historical and urban areas. *Adv. Geosci.,* Vol. 24, pp. 35–44.

Leucci G., Persico R. and Soldovieri F. (2007) Detection of fractures from GPR data: the case history of the Cathedral of Otranto. *J. Geophys. Eng.,* Vol. 4, pp. 452–461.

Lazzari M., Loperte A., and Perrone A. (2010). Near surface geophysics techniques and geomorphological approach. *Adv. Geosci.,* Vol. 24, pp. 35–44. to reconstruct the hazard cave map in historical and urban areas

Leone G. and Soldovieri F. (2003). Analysis of the distorted Born approximation for subsurface reconstruction: truncation and uncertainties effect", *IEEE Trans. Geoscience and Remote Sensing,* vol. 41, no. 1, pp. 66-74, Jan.

Loke, M.H and Baker R.D., (1996). Rapid least-squares inversion of apparent resistivity pseudosections by quasi-Newton method. *Geophys. Prospect.,* Vol. 44, pp.131-152.

Masini L, Persico R., Rizzo E, Calia A, Giannotta m. T., Quarta G., Pagliuca A. (2010). Integrated Techniques for Analysis and Monitoring of Historical Monuments: the case of S.Giovanni al Sepolcro in Brindisi (Southern Italy). Accepted for publication on Near Surface Geophysics.

McCann D. M., Forde M. C. (2001). Review of NDT methods in the assessment of concrete and masonry structures. *NDT&E International,* Vol. 34, pp. 71–84

Matta F., Bastianini F., Galati N., Casadei P., Nanni A. (2008). Distributed Strain Measurement in Steel Bridge with Fiber Optic Sensors: Validation through Diagnostic Load Test. *ASCE Journal of Performance of Constructed Facilities,* Vol. 22, pp. 264-273.

Measures R. M., *Structural monitoring with fiber optic technology.* Academic Press, 2001

Minardo A., Bernini R., Amato L., Zeni L., (2011). Bridge monitoring using Brillouin fiber-optic sensors. *IEEE Sensors Journal,* Vol. 99, pp.

Moustafa A., Mahadevan S., Daigle M., Biswas G. (2010). Structural and sensor damage identification using the bond graph approach. Structural Control and Health Monitoring, 17: 178-197.

Ohtsu M., Tokay M., Ohno K. and Isoda T., (2008). Elastic-wave methods for crack detection and damage evaluation in concrete. *Structural Faults & Repair-2008, 12th International Conference,* Edinburgh, UK: 10th -12th June 2008.

R. Persico, R. Bernini, Soldovieri F. (2005). The role of the measurement configuration in inverse scattering from buried objects under the Born approximation. *IEEE Trans. Antennas and Propagation*, Vol. 53, No.6, pp. 1875-1887.

Picozzi M., S. Parolai, M. Mucciarelli, C. Milkereit, D. Bindi, R. Ditommaso, M. Vona, M.R. Gallipoli, and J. Zschau. (2011). Interferometric Analysis of Strong Ground Motion for Structural Health Monitoring: The Example of the L'Aquila, Italy, Seismic Sequence of 2009. *Bulletin of the Seismological Society of America*, Vol. 101, No. 2, pp. 635–651, April 2011, DOI: 10.1785/0120100070.

Ponzo, F., Ditommaso, R., Auletta, G., Mossucca, A. (2010). A fast method for a structural health monitoring of Italian reinforced concrete strategic buildings. *Bulletin of Earthquake Engineering*. DOI: 10.1007/s10518-010-9194-6. Vol. 8, No. 6, pp. 1421-1434.

Ponzo F. C., Mossucca A., Di Cesare A., Nigro D., Dolce M., Moroni C. (2011). Seismic assessment of the R/C buildings: the case study of Di.Coma.C Centre after the L'Aquila (Italy) 2009 seismic sequence. *9th Pacific Conference on Earthquake Engineering. Building an Earthquake-Resilient Society.* 14-16 April, 2011, Auckland, New Zealand.

RILEM Recommendation NDT 1, *Testing of Concrete by the Ultrasonic Pulse Method*, Paris, December 1972.

Sasaky, Y. (1992). Resolution of resistivity tomography inferred from numerical simulation, Geophys. Prospect., Vol. 54, pp. 453-464.

A. Satriani, A. Loperte, M. Proto, and M. Bavusi (2011a). Building damage caused by tree roots: laboratory experiments of GPR and ERT survey. *Adv. Geosci.*, Vol. 24, pp. 133–137.

Satriani A., A. Loperte, M. Proto (2011b). Electrical resistivity tomography for coastal salt water intrusion characterization along the Ionian coast of Basilicata Region (Southern Italy). *Fifteenth International Water Technology Conference*, IWTC-15 2011, Alexandria, Egypt, 28-30 May.

Scott M,. Rezaizadeha A., Delahazab A., Santosc C. G., Moored M., Graybeale B. and Washerf G. (2003). A comparison of nondestructive evaluation methods for bridge deck assessment. *NDT&E Int.*, Vol., 36, pp. 245–55.

Seppänen A., Karhunen K., Lehikoinen A. & Kaipio J.P., (2009). *Electrical resistance tomography imaging of concrete.* Concrete Repair, Rehabilitation and Retrofitting II – pp. 571-577. Alexander et al (eds). © 2009 Taylor & Francis Group, London, ISBN 978-0-415-46850-3

Shaari A., Millard S. G. and Bungeyb J. H. (2004). Modelling the propagation of a radar signal through concrete as a low-pass filter. *NDT&E Int.*, Vol. 37, pp. 237–42.

Sharma, P.S. (1997). Enviromental and Engineering Geophysics. Cambridge *University Press.*

Shi B., Xu H., Chen B., Zhang D., Ding,Y., Cui H., Gao J.,(2003). Feasibility study on the application of fiber-optic distributed sensors for strain measurementin the Taiwan Strait Tunnel project. *Marine Georesources and Geotechnology,* Vol. 21, pp. 333-343.

Shaw M. R., Millard S. G. Molyneauxc T. C. K., Taylord M. J., Bungeyb J. H. (2005). Location of steel reinforcement in concrete using ground penetrating radar and neural networks. *NDT&E International*, Vol. 38, pp. 203–212.

Soldovieri F. and Orlando L.(2009). Novel tomographic based approach and processing strategies for multi-frequency antennas GPR measurements using multi-frequency antennas. *Journal of Cultural Heritage*, Vol. 10, pp. e83-e92.

Soldovieri F., Solimene R., (2010). Ground Penetrating Radar Subsurface Imaging of Buried Objects, *in Radar Technology*, IN-TECH, Vienna Austria, ISBN 978-953-307-029-2, Edited by: Guy Kouemou, January 2010.

Stolt, R. H. (1978) Migration by Fourier Transform. *Geophysics*, Vol. 43 pp. 23-48.

Tsoflias G.P., Van Gestelz J.P., Blankenship D.D. and Sen M., (2004). Vertical fracture detection by ex-ploiting the polarization properties of ground-penetrating radar signals. *Geophysics*, Vol. 69, pp. 803-810.

UNI EN 12504-1 (2002) Prove sul calcestruzzo nelle strutture - Carote - Prelievo, esame e prova di compressione

UNI EN 12504-2 (2001) Prove sul calcestruzzo nelle strutture - Prove non distruttive – Determinazione dell'indice sclerometrico

UNI EN 12504-3 (2005) Prove sul calcestruzzo nelle strutture - Parte 3: Determinazione della forza di estrazione

UNI EN 12504-4 (2005) Prove sul calcestruzzo nelle strutture - Parte 4: Determinazione della velocità di propagazione degli impulsi ultrasonici

Utsi V., Birtwisle A. and Coock J. (2008). Detection of subsurface reflective cracking using GPR. *Structural Faults & Repair*-2008: 12th Int. Conf., Edinburgh, UK, 10–12 June 2008.

Wenzel H. and Pichler D. (2005). *Ambient Vibration Monitoring*. Wilhey.

Zou L., Bao X., Wan Y., Chen L., (2005). Coherent probe-pump-based Brillouin sensor for centimetre-crack detection. *Opt. Let.*, Vol. 15, pp. 370-372.

7

Steel Building Assessment in Post-Earthquake Fire Environments with Fiber Optical Sensors

Genda Chen, Ying Huang and Hai Xiao
Missouri University of Science and Technology
Untied States of America

1. Introduction

This chapter is aimed to develop and calibrate a quasi-distributed optical sensor network of long period fiber gratings and extrinsic Fabry-Perot interferometers for simultaneous high temperature and large strain measurements, and to validate it for the post-earthquake assessment of a steel frame in high temperature environments. The steel frame represents typical steel buildings that are susceptible to the high temperature effect of earthquake-induced fires such as those observed at multiple locations during the March 11, 2011, Japan earthquake.

Critical buildings such as hospitals and police stations must remain functional for post-earthquake responses and evacuation immediately following a major earthquake event. However, they often experience large strains due to shaking effects as observed during recent earthquakes, causing permanent inelastic deformation. The post-earthquake fires associated with the earthquake-induced short fuse of electrical systems and leakage of gas devices can further strain the already damaged structures during the earthquakes, potentially leading to a progressive collapse of buildings. In a matter of seconds to an hour, tenants can be injured and trapped in the collapsed buildings, desperately waiting for rescue in helpless situations. Therefore, real time monitoring and assessment of the structural condition of critical buildings is of paramount importance to post-earthquake responses and evacuation in earthquake-prone regions. An accurate assessment of the buildings in these harsh conditions can assist fire fighters in their rescue efforts and save earthquake victims.

For structural condition assessment in fire environments, the most widely used, commercial sensing tools are based on electrical principles, including electrical resistance gauges for strain measurements and thermocouples for temperature measurements. Electrical resistance gauges were initially proposed in 1856 by Lord Kelvin (Thomson 1857). A strain gauge is basically a conductive metal foil that is printed on a non-conductive insulating flexible backing. It can be applied to the surface of a solid structure with suitable adhesives, such as cyanoacrylate. When perfectly attached, the foil and then the backing are deformed together with the structure to be monitored, causing a change in electrical resistance of the alloy. Due to electromechanical properties of the alloy, the foil and the adhesives, the maximum strain that a strain gauge can measure prior to its failure is typically limited to 1%. The readings from a strain gauge also vary with the temperature during measurements. They become significant in high temperature applications and must be compensated for temperature effects.

For strain measurements in high temperature environments, significant improvements on the temperature properties of strain gauges were made by Easerling (1963) and Gregory et al. (2007). Commercial products introduced by Vichy Micro-Measurement can be applied to measure a strain of *5,000 μɛ* at *1,000 °F*. However, wired strain gauges would likely lose their signals due to power outage during and immediately following a strong earthquake when structures being monitored are subjected to large strains and post-earthquake fire environments. For the same reason, conductive textiles that are operated with their electromechanical properties through a special design of "sensing string" will likely be malfunctional even though the sensors can be used for large strain measurements (Zhang 2006).

For temperature measurements, thermocouples were initially proposed in 1822 by Fourier and Oersted, and based on the thermoelectricity principle discovered by Thomas in 1821 (ASTM 1981). Since then, thermocouples have been widely used for high temperature measurements. Today, a variety of thermocouples are commercially available for various temperature ranges, including Type K, E, J, N, B, R, S, T, C, and M. A thermocouple is a device consisting of two different conductors (usually metal alloys) that produce a voltage proportional to a temperature difference between either ends of the pair of conductors. The change in voltage corresponds to the temperature to be measured. Like strain gauges, the measured voltage change and then the converted temperature may not be able to be transferred to a central data acquisition in post-earthquake fire environments.

The above review clearly indicates that a conventional monitoring system with strain gauges and thermocouples is inadequate for structural condition assessment in harsh environments, such as extremely high temperature in the event of a post-earthquake fire. On the other hand, an optical fiber monitoring system can provide a viable means for this application. In addition to its high temperature tolerance, optical fibers are compact, immune to electromagnetic interference, durable in acid environment, and capable of being integrated into various types of structures and materials. Once developed for large strain measurement capabilities, optical fiber sensors can be multiplexed and integrated into a quasi-distributed, multi-parameter sensing network in harsh environments.

Various fiber optical sensors have been developed mainly based on the intensity changes, gratings, and interferometers with the last two most widely investigated. Grating based optical sensors are led by the fiber Bragg grating (FBG) and the long period fiber grating (LPFG) technologies. A FBG sensor couples two light beams in their respective forward- and backward-propagating core-guided modes near a resonant wavelength, functioning like a wavelength-selective mirror (Othonos 1999). A LPFG sensor with a periodic refractive index perturbation of its fiber core has a period of the hundreds of micrometer and couples the core mode (guided light inside the core of the fiber) into the cladding modes at certain discrete wavelengths (known as resonance wavelengths) (Vengsarkar et al. 1996). For strain measurements at high temperatures, FBG sensors were applied by Wnuk et al. (2005) and Mateus et al. (2007); LPFG sensors were investigated by Huang et al. (2010a). In addition, extrinsic Fabry-Perot interferometer (EFPI) sensors were developed both for a small range of strain measurement (Xiao et al. 1997) and for a large range of strain measurement (Huang et al. 2010b, 2011). However, fiber optical sensors for large strain measurements at high temperature are yet to be studied. Furthermore, their applications for structural behaviour monitoring in post-earthquake fire environments have not been investigated.

In this chapter, an EFPI-based sensor is first introduced and characterized for large strain measurements with high resolution. In particular, the operational principle, signal processing algorithm, and experimental validation of the EFPI sensor are investigated in detail. The EFPI sensor developed for large strain measurements is then combined with a LPFG-based temperature sensor for simultaneous large strain and high temperature measurements. A straightforward signal decomposition technique is introduced for the evaluation of temperature and strain from the readings of an EFPI-LPFG sensor. Lastly, the developed fiber optical sensors are multiplexed in an optical fiber sensing network and validated in their application to a steel frame under a simulated post-earthquake fire environment for structural behaviour monitoring and assessment.

2. An EFPI based large strain sensor with high resolution

An EFPI sensor can be made by first inserting two cleaved optical fibers into a capillary tube and then bonding them to the tube with epoxy or in thermo fusion. This way of packaging improves the sensor's robustness in applications, but limits the sensor's dynamic range to the corresponding maximum deformation of the capillary tube. On the other hand, if the two cleaved ends are left unattached to the tube, the packaged device is essentially a displacement sensor. By dividing the measured displacement by the initial distance between the two cleaved ends, the device can be implemented as a large strain sensor. For example, if one or two ends of the fibers are adhered to a substructure to be monitored, the packaged device will experience an applied strain that can be determined from the measured displacement by the EFPI as the substructure deforms under an external load. In this case, the technological challenge remains in achieving high resolution during a large strain measurement. The concept of movable EFPI sensors was developed by Habel et al. (1996) and applied for strain measurements during the first few hours of hydration reaction in concrete (Habel et al. 1996, 1997, 2008).

Up to date, most applications of movable EFPI sensors deal with strain measurements in small range. The resolution of the sensors for large strain measurement has not been analysed systematically. Over the past twenty years, several data processing methods have been investigated to analyse data series from EFPI sensors in structural health monitoring applications (Liu et al. 2001 and Qi et al. 2003). Due to a small range of strain measurement, the so-called phase tracking method with relatively high resolution has been widely applied. Qi et al. (2003) developed a hybrid data processing method by combining several methods using a white light interferometer. Although only tested for a small dynamic range in strain measurements, the data processing technique appears to be a promising concept that can be extended to achieve an increasing dynamic range and resolution with EFPI sensors.

2.1 Operational principle
2.1.1 Sensor structure and measurement system

Fig. 1 shows the schematic of an EFPI sensor head and its associated measurement system. As illustrated by the enlarged view of the sensor head in Fig. 1(a), the EFPI is formed by two perpendicularly cleaved end faces of a single-mode optical fiber (Corning SMF-28). The left side of the fiber serves as a lead-in fiber and the right side serves as a low reflective mirror.

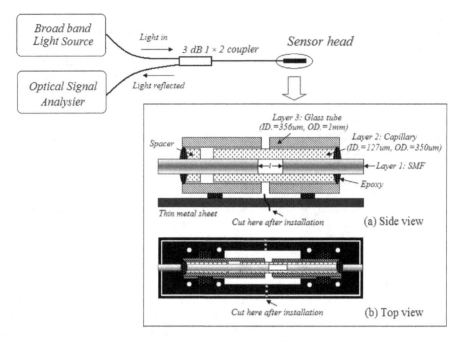

Fig. 1. Schematic of an EFPI sensor: sensor structure and measurement system

In theory, a Fabry-Perot cavity with a freely movable end face can be constructed by inserting two cleaved fiber ends into the two ends of a glass tube and gluing one side of the bare fiber to the tube. However, the freely movable bare fiber end, when not glued to the tube, is easy to break in applications since optical fibers are susceptible to any shear action. To address this issue, a three layer structure is used to package an EFPI strain sensor, including an inner, intermediate and outer layer. The inner layer or Layer 1 in Fig. 1(a) is an optical glass fiber of $125~\mu m$ in diameter. The intermediate layer or Layer 2 in Fig. 1(a) is a capillary glass tube with an inside diameter of $127~\mu m$ and an outside diameter of $350~\mu m$. The capillary tube is designed to guide the cleaved fiber to ensure that its two end faces can move in parallel. The outer layer or Layer 3 in Fig. 1(a) is a glass tube with an inside diameter of $356~\mu m$ and an outside diameter of $1,000~\mu m$, which is designed to enhance the overall stability of the packaged sensor. On one side (right) of the interferometer, all three layers are bonded together with epoxy as illustrated in Fig. 1(a). On the other side (left), the fiber is bonded to the third layer through an inserted spacer while the intermediate layer is unbounded to allow for free movement of the fiber end faces within the capillary tube.

The three-layer structure transfers a shear force from the bare fiber to the intermediate layer (spacer) during operation. As its diameter increases, the intermediate layer is less susceptible to the applied shear force so that the proposed structure can operate stably without breakage. The two pieces of the outer glass tube are bonded to a thin metal sheet both at the lead-in side and mirror side of the fiber sensor. The metal sheet has two perforated side strips and a pre-cut rectangular hole in the middle section. It can be bolted to a steel substructure in applications as illustrated in Fig. 1(b). The sensor installation is completed by cutting the two side strips across the perforation of the thin metal sheet. The separation of the attachment sheet ensures that the sensor actually measures the elongation of the steel substructure between the two sides of the cut section as clearly illustrated in the top view of Fig. 1(b). The distance between the two bolts closest to the cut section is defined as the gauge length of the sensor, which is $L = 2\ mm$ in this study. The EFPI cavity length is designated as l. Based on the proposed sensor structure, a sensor prototype was fabricated as shown in Fig. 2(a) and the micro-view of its sensor head can be seen in Fig. 2(b).

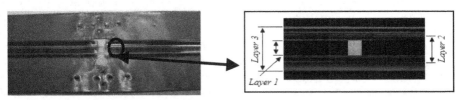

| (a) Top view of a sensor prototype: sensor head and metal attachment sheet | (b) Side view of the sensor head taken with an optical microscope |

Fig. 2. Sensor prototype

As shown in Fig. 1, the measurement system of an EFPI sensor uses a broadband light source (BBS) of wavelength from $1520\ nm$ to $1620\ nm$, which is generated by multiplexing a C-band (BBS 1550A-TS) and an L-band (HWT-BS-L-P-FC/UPC-B) Erbium Doped fiber amplified spontaneous emission (ASE). The light propagates into the EFPI sensor through a $3\ dB$ coupler. As the light travels through the lead-in fiber, part of the light is reflected at both cleaved end faces of the EFPI sensor, producing a backward travelling interference signal. The reflected interference spectrum coupled back by the coupler is detected by an optical spectrum analyzer (OSA, HP 70952B). A personal computer is used to record and process the interference spectra. Finally, the characteristic wavelength on the spectrum can be related to the cavity of the EFPI or the strain applied to the substructure in various resolutions with the signal processing algorithms presented in Section 2.1.2.

2.1.2 Signal processing algorithms

To simultaneously achieve a large dynamic range and high resolution in strain measurements, three data processing methods are introduced and studied to characterize their performance. These methods include 1) interference frequency tracking of the Fourier transform of a spectral interferogram, 2) period tracking and 3) phase tracking of the spectral interferogram.

Interference Frequency Tracking Method A low finesse EFPI can generally be modelled by a two-beam interference theory (Taylor 2008). The spectral interferogram of an EFPI typically represents a harmonic function of wavenumber with a dominant frequency known as the interference frequency. By taking the Fourier transform of such an interferogram, an approximate delta function of cavity length corresponding to the interference frequency is obtained (Liu et al. 2000). The cavity length of the EFPI, l, can then be calculated by Equation 1:

$$l = \frac{n\pi}{\left(v_E - v_S\right)}$$

(1)

in which v_S and v_E are the wavenumbers of the start and end points of an observation bandwidth, respectively, and n is an integer representing the Fourier series index.

It can easily be observed from Equation 1 that the minimum detectable cavity length change of an EFPI large strain sensor is $\pi/(v_E - v_S)$ when $n=1$. For a light source with a spectrum width of 100 nm, the detectable cavity length change is approximately 12 μm. This corresponds to strain resolution of approximately 6,000 $\mu\varepsilon$ when a gauge length of 2 mm is used. As indicated by Equation 1, the strain resolution is inversely proportional to the bandwidth of the light source. Higher resolution in strain measurement thus requires an optical source with a broader bandwidth, which can only be provided by a limited selection of equipment available in the market.

Period Tracking Method Due to the interrelation between period and interference frequency, the change in period of the spectral interferogram can also be used to determine the cavity length of an EFPI (Taylor 2008). The period of a spectral interferogram is defined as the distance between two consecutive valleys on the spectral interferogram. By introducing a wavenumber-wavelength relation ($v = 2\pi / \lambda$), the cavity length can be evaluated by Equation 2:

$$l = \frac{\lambda_1 \lambda_2}{2(\lambda_2 - \lambda_1)} \qquad (2)$$

where λ_1 and λ_2 ($\lambda_2 > \lambda_1$) represent the first and second wavelengths of two consecutive valleys on the spectral interferogram that can be directly taken from OSA measurements.

Let S_l be the resolution of a strain sensor, which is defined as the minimum detectable change in cavity length when using the period tracing method. Therefore, when λ_1 and λ_2 are assumed to be two independent random variables, S_l can be derived from Equation 2 and expressed into Equation 3:

$$S_l = \sqrt{(\frac{\partial l}{\partial \lambda_1})^2 (S_{\lambda_1})^2 + (\frac{\partial l}{\partial \lambda_2})^2 (S_{\lambda_2})^2} = \sqrt{\frac{\lambda_2^{\,4}}{4(\lambda_2 - \lambda_1)^4}(S_{\lambda_1})^2 + \frac{\lambda_1^{\,4}}{4(\lambda_2 - \lambda_1)^4}(S_{\lambda_2})^2} \qquad (3)$$

in which S_{λ_1} and S_{λ_2} represent the OSA measurement resolutions of the two consecutive valleys, respectively. Determined from the performance specifications of a particular OSA instrument, S_{λ_1} and S_{λ_2} are equal ($S_{\lambda_1}=S_{\lambda_2}=S_\lambda$) since the instrument has a consistent measurement resolution of wavelength within the specified observation bandwidth. In addition, within a relatively small observation spectrum range, both λ_1 and λ_2 can be approximated by the center wavelength of the range, λ_0. As a result, Equation 3 can be simplified into:

$$S_l \approx S_\lambda \frac{\lambda_0^{\,2}}{\sqrt{2}(\Delta\lambda)^2} \qquad (4)$$

where $\Delta\lambda$ is the wavelength difference between the two consecutive valleys. For the estimation of measurement errors, $\Delta\lambda$ at a given cavity length can be considered to be a constant within the wavelength bandwidth of observation, though $\Delta\lambda$ slightly increases with wavelength. Equations 2 and 4 indicate that the minimum detectable cavity length decreases in a quadratic manner as $\Delta\lambda$ increases or the cavity length decreases. In other words, as the cavity length increases, the resolution of cavity length or strain measurement becomes lower.

Phase Tracking Method Based on the two beam interference theory (Taylor 2008), the spectral interferogram reaches its minimum when the phase difference between the two beams satisfies the following condition:

$$\frac{4\pi l}{\lambda_v} = (2m+1)\pi \tag{5}$$

where m is an integer that can be estimated following the procedure as specified in Qi et al. (2003), and λ_v is the center wavelength of a specific interference valley. Taking the derivative of the cavity length with respect to λ_v yields

$$\frac{dl}{d\lambda_v} = \frac{2m+1}{4} \tag{6}$$

Therefore, the change in cavity length can be estimated from Equations 5 and 6 as follows:

$$\Delta l = \frac{\Delta\lambda_v}{\lambda_v} l \tag{7}$$

where $\Delta\lambda_v$ is the change in center wavelength of the specific interference valley and Δl is the change in cavity length. As Equation 7 indicates, the cavity length change is directly proportional to the wavelength shift of the interferogram and to the cavity length of the EFPI. Since the minimum $\Delta\lambda_v$ is represented by the instrument measurement resolution or $S_{\lambda 1}$ and $S_{\lambda 2}$, the resolution of the phase tracking method decreases linearly as the EFPI cavity length increases.

Comparison among Three Processing Methods Fig. 3 compares the theoretical strain measurement resolution of three data processing methods when $L=2$ mm. To cover a range of wavelength measurement resolution from various commercial OSAs, let $S_\lambda = \Delta\lambda_v$ be equal to 0.001 nm, 0.01 nm, and 0.1 nm. It can be clearly observed from Fig. 3 that the interference frequency tracking method has constant resolution of approximately $6,000$ $\mu\varepsilon$. The resolution of the period tracking method decreases in a quadratic manner as the EFPI cavity length increases. The resolution of the period tracking method is strongly influenced by the resolution of a particular OSA system. If $S_\lambda = \Delta\lambda_v = 0.01$ nm, the strain resolution of period tracking method is 600 $\mu\varepsilon$. If $S_\lambda = \Delta\lambda_v = 0.1$ nm, the period tracking method has higher resolution than the interference frequency tracking method for $l < 320$ μm. In addition, the resolution of the phase tracking method decreases linearly as the EFPI cavity length increases. Among the three methods, the phase tracking method has the highest resolution since it represents the local (most detailed information) change of phase. When $l = 320$ μm and the given OSA resolution is 0.1 nm, the strain resolution of the phase tracking method is 10 $\mu\varepsilon$ in comparison with $6,000$ $\mu\varepsilon$ for the other two methods. However, the phase tracking method can only measure a relatively small change of the cavity length within a 2π phase range to avoid ambiguity. Therefore, its operation range is limited to a change of approximately 0.75 μm in cavity length or a change of 375 $\mu\varepsilon$ in strain. On the other hand, the other two methods can be used to measure a large change of cavity length.

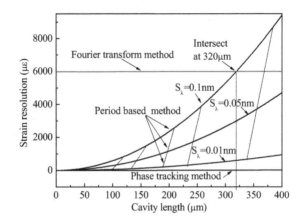

Fig. 3. Resolution as a function of cavity length

2.2 Experiment and discussions

To evaluate the performance of the proposed sensor for large strain measurements, an EFPI-based prototype sensor was constructed with transparent glass tubes so that any change in cavity length can be observed in the laboratory by using an optical microscope as shown in Fig. 2(b). The lead-in side of the fiber sensor was fixed on an aluminum block and the fiber mirror side of the sensor was attached to a computer-controlled precision stage so that the cavity length can be controlled precisely. The gauge length of the strain sensor was set to 2 mm. The reference strain, which will be further discussed in Fig. 5, was determined by dividing the change in cavity length, directly measured by stage movement, by the gauge length. The strain detected by the EFPI sensor is obtained by dividing the cavity length calculated from an EFPI signal to the gauge length.

Fig. 4(a) presents two interferograms of the EFPI sensor prototype with a cavity length of 65 μm and 175 μm, respectively. It can be observed from Fig. 4(a) that the interference frequency increases as the EFPI cavity length increases or as more fringes are condensed into a given observation spectrum range. However, the range of interference signal intensities decreases as the EFPI cavity length increases. The signal range is often quantified by a fringe visibility (V) as defined by Equation 8:

$$V = \frac{I_{\max} - I_{\min}}{I_{\max} + I_{\min}} \tag{8}$$

where $I_{\max}$ and $I_{\min}$ represent the maximum and minimum intensities of an interference signal, respectively. The fringe visibility determined from Equation 8 is plotted in Fig. 4(b) as a function of cavity length. The experiment stopped when the fringe visibility dropped below 20%, corresponding to a maximum cavity length of approximately 265 μm of the prototype sensor. The drop in fringe visibility as a function of cavity length was mainly caused by the divergence of the output beam from the lead-in fiber, which was governed by the numerical aperture (NA) of the fiber (Han et al. 2004). Other potential factors such as misalignment are negligible in this study since the glass tubes of the three-layer sensor prototype were assembled with a tight tolerance.

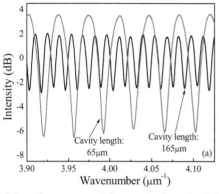

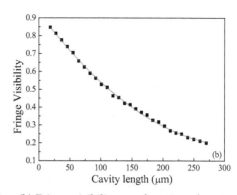

(a) Interferograms with a cavity length of 65 μm and 175 μm

(b) Fringe visibility as a function of cavity length

Fig. 4. Interferogram and fringe visibility

To investigate the measurement resolution of the interference frequency tracking method, large strain measurement experiments were designed and executed. During various tests, the cavity length of the sensor prototype ranged from *15 μm* to *265 μm* at *10 μm* intervals. The maximum change of cavity length was approximately *250 μm*, corresponding to a dynamic strain range of *12.5%*. Fig. 5 relates the reference strain measured by the change in stage movement to the strain measured by the change in cavity length of the EFPI sensor. The theoretic values were directly calculated based on the stage movement; they follow a straight line with a slope of *1:1* as represented by the solid line in Fig. 5. The experimental data points demonstrated only slight fluctuations around the theoretic line. To compare the measurement resolution of the interference frequency tracking method and the period tracking method, refined experiments were conducted within a strain range of *11,000 με* to *21,000 με*. In this case, the precision stage was moved at *2 μm* intervals, giving rise to a strain change of *1,000 με* between two consecutive measurements. The results from the refined experiments processed with both the interference frequency tracking and period tracking methods are presented as an inset in Fig. 5. It can be seen from Fig. 5 that the theoretical prediction strongly agrees with the test data points that were processed with the period tracking method and the measured strains processed with the interference frequency tracking method follow a zig-zag trend with respect to the theoretic prediction. This comparison indicates that the interference frequency tracking method is unable to resolve a strain difference within an interval of *12 μm* in cavity length. This length resolution corresponds to a strain measurement of approximately *6,000 με*, which agrees with the calculated strain resolution that is limited by the light source bandwidth of *100 nm*.

The relative accuracy between the interference frequency method and period tracking method is supported by Fig. 3 since the cavity length observed during the refined experiments was significantly less than *320 μm* when the two methods had the same resolution. The interference frequency tracking method is advantageous over the other two methods in terms of computational efficiency and constant resolution over the entire dynamic range. In addition, it is immune to localized spectrum distortions that could potentially result in large errors when waveform based signal processing methods are used.

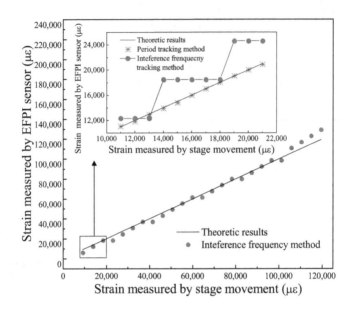

Fig. 5. Measured stains processed with the interference frequency tracking method (inset: comparison between the frequency and period tracking methods)

To verify the accuracy of the phase tracking method, more refined experiments were performed with a smaller stage movement interval of *0.1 μm*. The cavity length of the EFPI sensor was set to range from *15 μm* to 30 μm, which corresponds to a strain of *7,500 με*. Fig. 6(a) shows two representative spectral interferograms of the EFPI sensor at two consecutive stage positions with a cavity length difference of *0.1 μm*. Fig. 6(b) compares the measured strains processed with the phase tracking and the period tracking methods. It can be observed from Fig. 6(b) that the theoretically predicted strain is in agreement with the strain data points processed with the phase tracking method and that of the period tracking method shows a notable deviation from the theoretic prediction based on the reference strains. This comparison indicates that refined resolution can be achieved with the use of the phase tracking method. The maximum deviation of the period tracking method was estimated to be *50 με* at an EFPI cavity length of *30 μm*, which is consistent with the theoretic prediction given in Fig. 3. The deviation is expected to further increase as the EFPI cavity length increases. However, it is worth repeating that the period tracking method can measure a large range of strain while the phase tracking method is limited to a strain measurement range of approximately *375 με*, which corresponds to a phase shift of 2π.

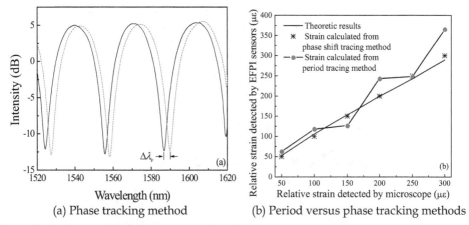

(a) Phase tracking method

(b) Period versus phase tracking methods

Fig. 6. Typical spectral Interferograms and strain measurements

3. A quasi-distributed optical fiber sensing system for large strain and high temperature measurements

A hybrid LPFG/EFPI sensor was designed and fabricated by linking a CO_2 laser induced LPFG with a movable EFPI in series. On one hand, LPFG is at least two orders (approximately 100 times) more sensitive to temperature than strain, and thus regarded as a temperature sensing component of the hybrid sensor. On the other hand, packaged with a glass tube only, EFPI has a weak cross effect of temperature on strain measurements in the order of $0.5~\mu\varepsilon/^{\circ}C$, depending upon thermal coefficients of the optical fiber and the glass tube. Therefore, EFPI mainly works as a strain sensing component of the hybrid sensor (Huang et al. 2010).

3.1 Sensor structure and measurement system

Built upon Fig. 1(a), Fig. 7 shows the schematic of a hybrid fiber optical LPFG/EFPI sensor structure. The EFPI was formed by two perpendicularly cleaved end faces of a single-mode optical fiber (Corning SMF-28). The left optical fiber with written gratings (LPFG) serves as a lead-in fiber and the right cleaved end face functions like a low reflective mirror. Both the cleaved end faces were placed inside a capillary tube with an inner diameter of $127~\mu m$ and an outer diameter of $250~\mu m$ to ensure the two end faces move in parallel. One end of the fiber was bonded on the capillary tube and the other end was not. The sensor installation was completed by bonding the capillary tube and the right fiber to a host structure over a gauge length of L between the two bonding points. In this study, $L = 2~mm$. The EFPI cavity length was designated as l. The bonding between all the components was completed after high temperature tolerable adhesives had been applied (Huang et al. 2008).

Fig. 7 also illustrates the expected main mechanism of light reflection and transmission around a hybrid sensor. As light at the light-in end of the fiber transmits through the LPFG, one part of the light is reflected by the left cleaved end face and the other part is reflected by the right cleaved end face. Since the LPFG is very close to the EFPI, typically less than 5 cm apart, both the EFPI-reflected light beams will be further reflected by the LPFG mirror towards the EFPI. At the light-out end of the fiber, the two branches of secondly reflected lights by the LPFG mirror thus form an interferometer at the OSA output spectrum together with the spectrum of the LPFG itself. The spatial frequency of the interferometer is only a function of the EFPI cavity length. The refractive index change of the interferometer does not affect the transmission signal of the LPFG itself as a sensor component.

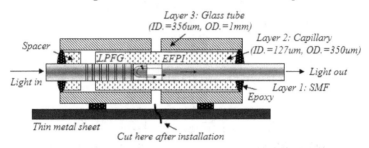

Fig. 7. Sensor structure and operational principle of a hybrid EFPI/LPFG sensor

Fig. 8 shows an optical sensing system of multiple hybrid LPFG/EFPI sensors. Light from a broadband source was channelled through each sensor head by two N-channel D configuration Input/output optic switchers, and received by the OSA. The OSA was then connected to a personal computer for real time data processing. A typical transmission spectrum seen on the OSA is illustrated in Fig. 8 as well. With the two N-channel D configuration optic switchers, N strategic locations of a critical structure can be monitored for simultaneous large strain and high temperature measurements.

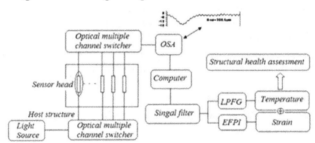

Fig. 8. An optical sensing system of hybrid EFPI/LPFG sensors

3.2 Decomposition of the signal from a hybrid LPFG/EFPI sensor

To enable the decomposition of a recorded signal into components from each LPFG sensor and each EFPI sensor in the hybrid sensing system, LPFG sensors must be designed with different wavelengths for multiplexing and EFPI sensors are given different optical paths. The signal processing algorithm for the signal decomposition is discussed below.

To measure temperature and strain at a specific location of a structure, the distance between the LPFG and EFPI components of a hybrid sensing system must be short, i.e., within 5 cm. With such a short distance between the two sensor components, the transmission signal of a hybrid sensor approximately represents a combined effect of individual LPFG and EFPI components. Fig. 9(a) illustrates a typical spectral interferogram of the hybrid LPFG/EFPI sensor as detailed in Fig. 7 with a cavity length of over 20 μm, and Fig. 9(b) shows the phase and amplitude of the Fast Fourier Transform (FFT) of the interferogram. It can be seen from Fig. 9(b) that the spatial frequency of the LPFG component is low and that of the EFPI

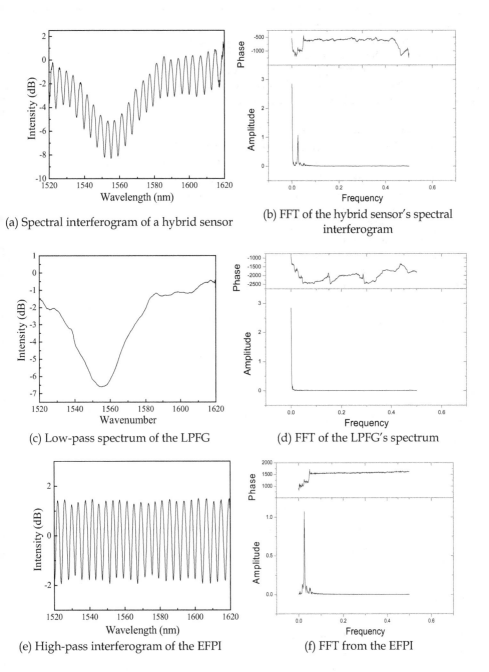

(a) Spectral interferogram of a hybrid sensor

(b) FFT of the hybrid sensor's spectral interferogram

(c) Low-pass spectrum of the LPFG

(d) FFT of the LPFG's spectrum

(e) High-pass interferogram of the EFPI

(f) FFT from the EFPI

Fig. 9. Transmission spectra and their FFTs of various sensors with low-pass and high-pass filters

component is significantly higher. Therefore, to decompose the signal into two parts from the LPFG and EFPI components, respectively, a low-pass filter was applied. Figs. 9(c, d) present the spectrum of the LPFG, a low-passed signal from Fig. 9(a), and its Fourier transform. As such, the interferogram of the EFPI was then obtained by subtracting Fig. 9(c) from Fig. 9(a), as shown in Fig. 9(e). Its Fourier transform is presented in Fig. 9(f). The cavity length of the EFPI can be calculated from the decomposed signal based on the spectral frequency tracking method as discussed in Section 2.1.2.

3.3 Experiments and discussions

As indicated in Fig. 5, the EFPI component only of a hybrid LPFG/EFPI sensor can measure a gap distance change of up to 260 μm, corresponding to a strain of 12% for a gauge length of 2 mm, and temperature as high as 1292 °F (700 °C). In this section, a hybrid LPFG/EFPI sensor was tested in the laboratory for simultaneous strain and temperature measurements. The hybrid sensor was installed on two steel channels of 2 mm apart, which were fixed on two computer controlled precise stages. The steel channels together with the hybrid sensor were further put into the high temperature furnace (made by Thermo Electron Corporation). Figs. 10(a, b) illustrate the measured temperature and strain. The hybrid LPFG/EFPI worked well up to 1292 °F (700 °C) and the difference between the strain calculated from the interference frequency method and that from the movable stage was within 5%, which is acceptable for large strain measurement. Fig. 10(b) also illustrates that the influence of temperature on the EFPI signal was small and insignificant.

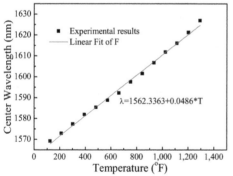 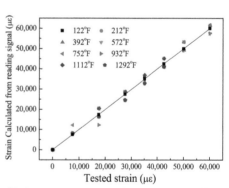

(a) Wavelength change with temperature   (b) Strain accuracy and temperature effect

Fig. 10. Temperature sensitivity, cross temperature-strain effect, and strain measurement accuracy of a LPFG/EFPI sensor

4. Post-earthquake assessment of steel buildings under simulated earthquake and fire loadings with an optical fiber sensor network

4.1 Design of an idealized steel frame

A single-bay rigid frame of one top beam and two columns is considered in this study. The frame was made of A36 steel and built with hot-rolled S-shapes as shown in Fig. 11. To illustrate a potential switch of failure modes from one column to another under earthquake and post-earthquake fire loadings, respectively, a substructure of the frame consisting of one column (#1 in Fig. 11) and the top beam was tested under a static lateral load to simulate earthquake effects and the entire frame with two identical columns was tested with the other column (#2 in Fig. 11) placed in a high temperature environment to represent post-earthquake fire effects. The former is referred to as an L-shaped frame and the latter is referred to as a two-column frame for clarity in the following discussions.

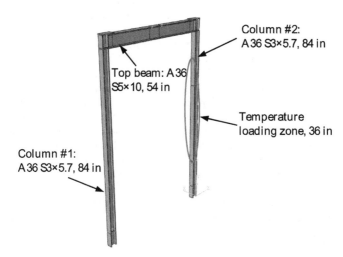

Fig. 11. Rendering of the steel frame

The dimensions of the steel columns were determined based on the size of an electric furnace (Lindberg/Blue M Tube Furnaces) used to simulate the high temperature effect of post-earthquake fires on the progressive collapse of the frame structure. The overall dimension of the furnace used for testing is 22×54×16 in. (55.9×137.2×66 cm) with an actual heating zone of 36 in. (91.4 cm) in length and an inner diameter of 6 in. (15.24 cm). Considering additional spaces required for the assembling (welding of stiffeners, beam-column joint, and column-tube connection) of the two-column frame after column #2 has been placed through the round furnace, the length of the columns was selected to be 84 in. (213.4 cm). To provide a sufficient space for frame deformation at high temperature, the columns of the steel frame were selected as S3×5.7, which has a flange width of 3 in. (7.62 cm) and a height of 4 in. (10.16 cm). To design a rigid beam, the stiffness of the top beam was set at least 5 times that of each column. As such, a hot-rolled S5×10 beam was selected. Since the anchors on the strong floor in the structures laboratory are spaced 36 in. (91.44 cm) apart, the length of the top beam was selected to be 56 in. (142.2 cm), which is equal to the anchor spacing plus the width of one bolted plate on the floor and twice the width of an angle stiffener. To ensure a rigid beam-column connection, three stiffeners were provided on each column: a 3×12×0.5 in. (7.62×30.48×1.27 cm) stiffener placed on the top cross section of the column, and two 1.4×3.9×0.5 in. (3.556×9.9×1.27 cm) stiffeners placed between the two flanges of the column on two sides of the column web, extending the bottom flange of the beam. Stiffeners were welded to the steel frame at the beam-column connection. The overall design of the steel frame is shown in Fig. 11. The column subjected to earthquake effects only is referred to as Column #1 while the other column directly subjected to earthquake-induced fire effects is referred to as Column #2.

4.2 L-shaped steel frame and earthquake-induced damage
4.2.1 Test setup and instrumentation under lateral loading

To simulate earthquake damage of the steel frame (Column #1 only), Column #1 and the top beam was placed on the strong floor and subjected to cyclic loading. Fig. 12 shows the test setup of the L-shaped frame and its prototype in the inset. The L-shaped frame was welded on a steel tube of *6×6×132 in. (15.24×15.24×335 cm)* with ½ in. *(1.27 cm)* wall thickness. In addition, two triangle brackets were individually welded to the two flanges of the column and the square tube to ensure a rigid connection between the column and the tube. The square tube was anchored into the strong floor at four anchor locations. To prevent it from experiencing large deformation, the square tube was stiffened near the base of the column by three stiffener plates of *12×5.5×0.5 in. (30.48×13.94×1.27 cm)*. The stiffeners were welded on the three sides of the square tube: column base face and two side faces. To approximately represent the two-column frame behavior, the free end of the top beam was transversely restrained by a roller-type support. A cyclic load was applied to the top of the column or the top beam by means of a hydraulic actuator. The applied load was measured by a 5-kip load cell installed between the actuator and the frame. To monitor the structural behavior under the cyclic load, *15* conventional strain gauges were deployed and distributed along the column and the beam as located in Fig. 12. They are designated with a prefix of SG#. For example, SG#1 means the strain gauge #1 that was deployed to monitor the strain in the column-to-tube connection. Similarly, SG#12 was used to assess the beam-column joint condition. In addition, two linear variable differential transformers (LVDTs) were respectively deployed *8 in. (203.2 mm)* above the column base and *14 in. (355.6 mm)* below the bottom flange of the top beam. LVDT#1 was deployed at the bottom of the column to ensure that the column is not displaced during testing. LVDT#2 was deployed at this location for convenience.

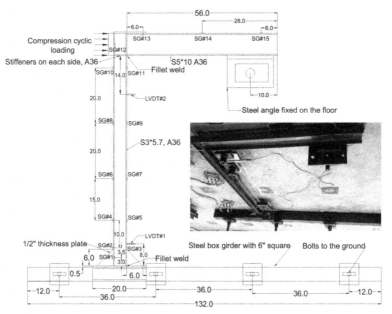

Fig. 12. Test setup and instrumentation of the L-shaped steel frame with a prototype inset (unit: in.)

4.2.2 Loading protocol and simulated earthquake damage

Fig. 13 presents the cyclic loading protocol, measured strains and displacements of the L-shaped frame structure. As shown in Fig. 13(a), five cycles of loading were applied to the frame following a sawteeth pattern. The first four cycles of loading reached *3.4 kips (15.1 kN)* at which the column expects to experience initial yielding, and the last cycle reached *3.8 kips (16.9 kN)* to ensure that the column is subject to inelastic deformation. For all cycles, the frame structure was loaded and then unloaded at a rate of *-5.46 lb/sec (-24.29 N/sec)* and *-10.92 lb/sec (-48.57 N/sec)*, respectively.

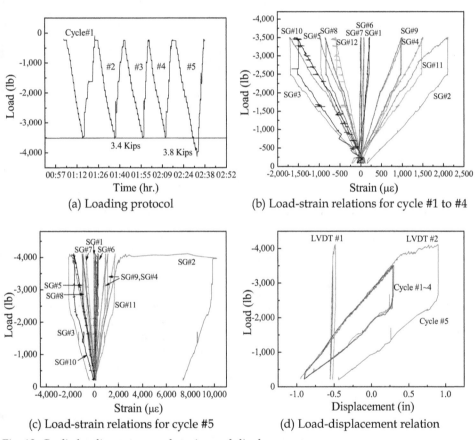

Fig. 13. Cyclic loading, measured strains and displacements

The measured strains (raw data) of the column are presented as a function of the applied load in Fig. 13(b) during the first four cycles of loading and in Fig. 13(c) during the last cycle. For the first four loading cycles, the maximum strain at the bottom of the column (SG#2) was approximately 0.2%, indicating initial yielding of the test frame. For the fifth cycle, the strain reached 1% as the load was held at *3.8 kips (16.9 kN)* for a few seconds. After unloading, a permanent strain of 0.75% remained at the column base (SG#2). Throughout the tests, the maximum strains in the beam-column and column-tube connection areas are both insignificant due to their significantly stronger designs than that of the column member. The maximum strains at locations slightly away from the connection areas are the highest as shown in Fig. 14 for strain distributions along the height of the column (outside face) during the first four cycles and the fifth cycle of loading. It can be clearly observed from Fig. 14 that Column #1 was subject to double curvatures with a zero strain around *20 in. (50.8 cm)* above the column base. The extent of inelastic deformation was mainly limited to the bottom portion of the column during the fifth cycle.

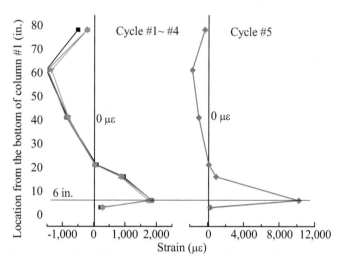

Fig. 14. Distribution of the maximum strains along the column height

The displacement change with load is presented in Fig. 13(d) in various loading cycles. It can be observed from Fig. 13(d) that the displacement change from LVDT#1 is negligible, indicating that the frame base was basically fixed to the strong floor. The largest displacement of *1.8 in. (4.572 cm)* was observed at the top of the column (location of LVDT#2), leaving behind *0.5 in. (1.27 cm)* permanent deformation in the column when unloaded. The permanent plastic strain and deformation introduced by the cyclic loading represented a large strain condition that can be induced by an earthquake event.

4.3 Two-column steel frame and fire induced damage
4.3.1 Test setup and instrumentation under fire loading

Fig. 15 shows the overall test setup of the two-column frame and various data acquisition systems used during tests as well as side views of the frame (shown in the inset). The frame was fixed to the strong floor at the base of both columns through the same square tube as used for cyclic tests. To simulate the gravity effect, the frame is subjected to a vertical load applied in displacement control by a hydraulic actuator against a rigid reaction beam. One column (#2) passed through a high temperature furnace before it was welded to the remaining L-shaped frame. A wood frame was built as a lateral support for instrumentation. Commercial sensors included 10 K-type thermocouples (TM), 5 high temperature strain gauges (HSG), and 15 strain gauges deployed during the earthquake test. Fiber optical sensors consisted of 1 FBG temperature sensor, 2 LPFG high temperature sensors, 5 EFPI large strain sensors, and 2 hybrid EFPI/LPFG sensors. As shown in Fig. 16, to protect them from potential damage, the fiber optical sensors on Column #2 were attached to the inside face of its flanges. The FBG and 1 LPFG were placed immediately above the furnace to monitor the temperature and strain at this transition area. The other LPFG sensor was located at the base of Column #1 to ensure a good estimate on the boundary condition. Three out of the five EFPI sensors were placed at the end of the heating zone (36″ total height) and two were located at the third and two-third points as illustrated in Fig. 16. One hybrid EFPI/LPFG sensor was placed at the bottom of the furnace and the other was at the 2/3 length from the bottom of the furnace together with one EFPI sensor.

To closely monitor the structural behavior of the entire steel frame, other than fiber optical sensors, a comprehensive sensing network of commercial sensors was also developed and applied to the frame structure as illustrated in Fig. 17. As illustrated in an inset in Fig. 17, ceramic adhesives that can endure high temperature up to 2012 °F were used to attach optical sensors on the inside surface of the column flanges.

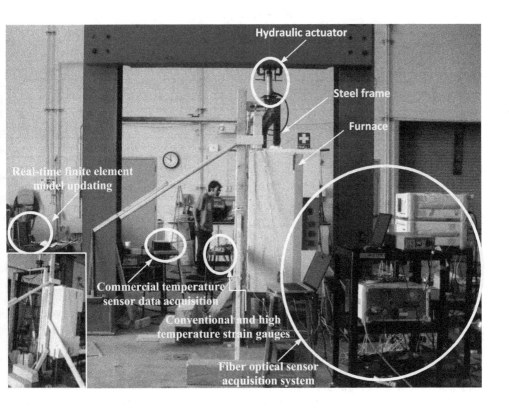

Fig. 15. Overview of the two-column frame test setup and data acquisition systems

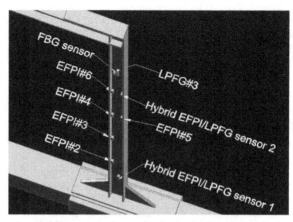

Fig. 16. Specific locations of fiber optical sensors in three-dimensional view

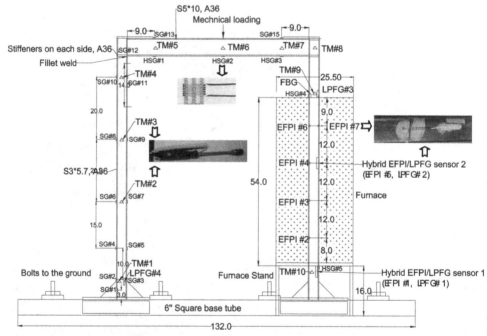

Fig. 17. Instrumentation for simulated post-earthquake fire tests with photos showing the EFPI and hybrid EFPI/LPFG sensors, high temperature strain gauges, and thermocouples

4.3.2 Loading protocol and simulated fire damage

A Lindberg/Blue M tube furnace made by Thermo Scientific was used to provide a high temperature environment that simulates the high temperature effect of a post-earthquake fire. It had three temperature zones that can be programmed and operated independently. In this study, the three temperature zones were programmed to have the same temperature increase profile. As shown in Fig. 18(a), temperature was increased at a rate of *18 °F/min* from *72 °F* (room temperature, *22 °C*) to *1472°F* (*800 °C*) by an interval of *180 °F* (*100 °C*) or *90 °F* (*50 °C*). At each temperature level, the test paused for *10* minutes to ensure that the temperature distribution is stable both inside and outside the furnace.

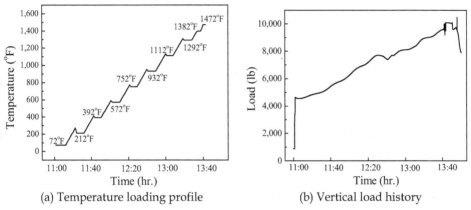

(a) Temperature loading profile   (b) Vertical load history

Fig. 18. Loading condition during post-earthquake fire tests

As temperature increased, Column #2 was elongated, increasing the stroke of the hydraulic actuator. As a result, the vertical load applied on the top beam by the displacement-controlled actuator was increased significantly. Corresponding to the increasing of furnace temperature, Fig. 18(b) shows that the frame was subjected to a vertical load of *4.6 kips* (*20.46 kN*) to *10 kips* (*44.48 kN*). Since the applied load was introduced to mainly simulate the gravity effect on the frame structure, an alternate scheme to load the frame needs to be developed in the future to maintain a nearly constant gravity effect.

Fig. 19(a) shows the measured temperatures from *4* LPFG sensors. LPFG#2, which was placed inside the furnace, measured the furnace temperature profile as shown in Fig. 18(a). The temperature on the top of the furnace increased from room temperature to *550 °F* as indicated by the measurement of LPFG#3. Fig. 19(b) shows the measured strains from 7 EFPI sensors after temperature compensation, including those as part of the two hybrid sensors. Note that EFPI#6 and EFPI#7 were malfunctional after strains exceeded *11%*. This is because all EFPI sensors were designed to have the maximum strain of approximately *11%* with an initial cavity of *50~60 μm*. At the top of the heating zone, the steel column is subject to a strain of *8%* at *752 °F* and would have been strained for over *11%* at *927 °F* based on the strain increasing trend over the time. Other locations with maximum strains of less than *11%* were continuously monitored by the EFPI sensors till the end of high temperature tests. It is also observed from Fig. 19(b) that the strains measured by EFPI#4 and EFPI#5 are similar, both slightly larger than that of EFPT#3 for most part. These three sensors were all located inside the furnace. However, the strain measurements by EFPI#2 (bottom of the furnace) were substantially smaller than those from EFPI#6 and EFPI#7 (top of the furnace) even though they were both located at the end of the furnace. This comparison indicated the effect of gravity on the temperature distribution at the boundary of the furnace.

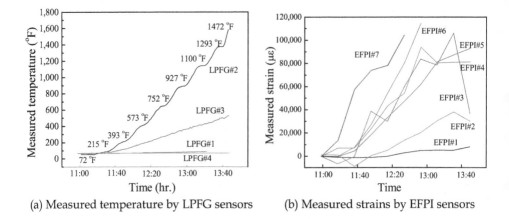

(a) Measured temperature by LPFG sensors   (b) Measured strains by EFPI sensors

Fig. 19. Measurements from LPFG and EFPI sensors

By comparing Fig. 13(c) with Fig. 19(b), relative effects of simulated earthquakes and earthquake-induced fires on column stability of the test frame can be discussed. During the cyclic load tests simulating earthquake effects, Column #1 was subjected to approximately 1% strain or ductility of 8 for A36 steel. This level of strain likely represents the effect of a moderate earthquake. In a post-earthquake fire environment, Column #2 was subjected to over 10% strain or ductility of over 80 for A36 steel. At 1472 °F, Column #2 became unstable due to extensive strain and deformation, resulting in a progressive collapse of the steel frame. This was confirmed by a visual inspection of the tested frame towards the completion of the experiment. It can thus be concluded that the fire-induced inelastic deformation of steel structures can far exceed the earthquake-induced deformation under a moderate earthquake event.

To understand how temperature affects the structural condition of the frame outside the furnace, Fig. 20 presents the strain measurements from strain gauges and high temperature strain gauges as the furnace temperature increased. Without direct temperature loading on Column #1, the permanent plastic strain of 0.75% induced by the simulated earthquake (SG#2) remained nearly constant during the simulated post-earthquake fire condition. The strains at other locations varied little with temperature as well.

4.3.3 Validation of the optical fiber sensing network against conventional sensors

The strains measured by EFPI#1 and HSG#5 are compared in Fig. 21(a) near the bottom of Column #2 immediately below the furnace. The two measurements showed a similar trend with a correlation coefficient of 0.963. This comparison verified the viability of using fiber optical sensors for strain measurements. Similarly, Fig. 21(b) compares various temperature measurements by TM#9, LPFG#3, and FBG sensors near the top of Column #2 immediately above the furnace. Overall they are in good agreement even though the LPFG sensor appeared to give a better comparison with the thermocouple in two temperature ranges as seen in Fig. 21(b).

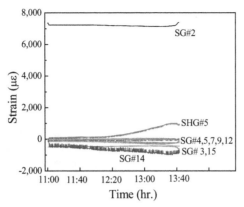

Fig. 20. Change of strains in the frame outside the furnace

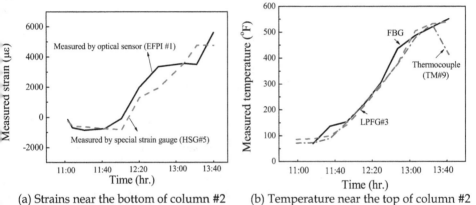

(a) Strains near the bottom of column #2   (b) Temperature near the top of column #2

Fig. 21. Comparison between fiber optical sensors and commercial sensors

5. Conclusions

In this chapter, an optical fiber sensing network of extrinsic Fabry-Perot interferometers and long-period fiber gratings sensors has been developed and validated with laboratory experiments for large strain, high resolution measurements in high temperature environments. The operational principle of the hybrid sensors and three signal processing algorithms were presented, including interference frequency tracking, period tracking, and phase tracking methods. A prototype of the hybrid sensors has achieved strain resolution of 10 $\mu\varepsilon$ within a 12% dynamic range at temperature up to 1472 °F (800 °C). Through extensive tests on a steel frame in a high temperature environment, the developed fiber optical sensors were validated against commercial temperature and strain sensors in their limited measurement ranges; the optical fiber sensing network was proven effective to monitor the structural condition of the frame in real time. On one hand, one column of the two-column frame structure was subjected to significant inelastic deformation during an earthquake event. On the other hand, the other column of the steel structure may experience extensive strains under post-earthquake fires, resulting in a progressive collapse of the structure. Overall, the developed optical sensing network and fiber optical sensors have been demonstrated to be viable devices for the monitoring and behaviour assessment of steel structures under extreme loads such as earthquakes and earthquake-induced fires. They can be employed in practical applications under harsh conditions.

6. Acknowledgments

Financial support to complete this study was provided in part by the U.S. National Science Foundation under Award No.CMMI-0825942 and by the Mid-America Transportation Centre under Award No. 0018358. The findings and opinions expressed in this chapter are those of the authors only and do not necessarily reflect the views of the sponsors.

7. References

ASTM Committee E20. (1981). *Manual on the Use of Thermocouples in Temperature Measurement (STP-470B)*, pp.28-36, ISBN: 0-8031-0502-9, West Conshohocken: ASTM International, 1981

Bhatia, V. & Vengsarkar, A. (1996). Optical fibre long period grating sensors, *Opt. Lett.*, Vol. 21, No. 9, (March 1996), pp.692-694.

Corte, G. D.; Landolfo, R. & Mazzolania, F. M. (2003). Post-earthquake fire resistance of moment resisting steel frames, *Fire Safety Journal*, Vol. 38, No.7, (November 2003), pp. 593-612.

Easerling, K. E. (1963). High temperature resistance strain gauges, *Brit. J. Appl. Phys.*, Vol. 14,No. 2, (February 1963), pp.79-84.

Gregory, O. J. & Chen, X. M. (2007). A low TCR nano-composite strain gage for high temperature aerospace applications, *IEEE Sensors 2007 Conference*, pp.624-627, ISSN 1930-0395, Atlanta, GA, USA, Oct. 28-31 2007.

Habel, W. R.; Hofmann, D.; Hillemeier, B. & Basedau, F. (1996). Fibre sensors for damage detection on large structures and for assessment if deformation behavior of cementitious materials, *Proc. of 11th Engineering Mechanics ASCE-Conference,* pp.355-358, Fort Lauderdale, FL, USA., 19-22 May, 1996.

Habel, W. R. & Hillemeier, B. (1998). Non-reactive measurement of mortar deformation at very early ages by means of embedded compliant fibre-optic micro strain sensors, *Proc. of 12th Engineering Mechanics ASCE-Conference,* pp.799-802, La Jolla, CA, USA., 17-20 May 1998.

Habel, W. R.; Hofmann, D. & Hillemeier, B. (1997). Deformation measurement of mortars at early ages and of large concrete components on site by means of embedded fibre-optic microstrain sensors, *Cement and Concrete Composites,* Vol.19, No.1, (May 1998), pp.81-102.

Han, M. & Wang, A. (2004). Exact analysis of low-finesse multimode fibre extrinsic Fabry-Perot interferometers, *Applied Optics*, Vol.43, No. 24, (August 2004), pp.4659-4666.

Huang, Y.; Chen, G.; Xiao, H.; Zhang, Y.; & Zhou, Z. (2011). A quasi-distributed optical fibre sensor network for large strain and high-temperature measurements of structures, *Proc. SPIE,* Vol. 7983, (March 2011), pp.17-27.

Huang, Y.; Wei, T.; Zhou, Z.; Zhang, Y.; Chen, G. & Xiao, H. (2010b). An extrinsic Fabry–Perot interferometer-based large strain sensor with high resolution, *Meas. Sci. Technol.,* Vol. 21, No. 10, (Sep. 2010), pp.105308.1-8.

Huang, Y.; Zhou, Z.; Zhang, Y.; Chen, G. & Xiao, H. (2010a). A temperature self-compensated LPFG sensor for large strain measurements at high temperature, *IEEE Trans. Instru. & Meas.,* Vol. 50, No.11, (Nov. 2010), pp. 2997 – 3004.

Li, Y. J.; Wei, T.; Montoya, J. A.; Saini, S. V.; Lan, X. W.; Tang, X. L.; Dong, J. H. & Xiao, H. (2008). Measurement of CO_2-laser-irradiation-induced refractive index modulation in single-mode fibre toward long- period fibre grating design and fabrication, *Applied Opt.,* Vol. 47, No.29, (Oct. 2008), pp. 5296-5304.

Liu, T. & Fernando, G. F. (2000). A frequency division multiplexed low-finesse fibre optic Fabry–Perot sensor system for strain and displacement measurements, *Review of Scientific Instruments,* Vol.71, No.3, (Nov. 1999), pp.1275-1278.

Mateus, C. F. R. & Barbosa, C. L. (2007). Harsh environment temperature and strain sensor using tunable VCSEL and multiple fibre Bragg gratings, *2007 SBMO/IEEE MTT-S International Microwave & Optoelectronics Conference,* pp.496-498, ISBN: 978-1-4244-0661-6, Brazil, Oct. 29 -Nov. 1 2007

Othonos, A. & Kalli, K. (1999). *Fibre Bragg Gratings: Fundamentals and Applications in Telecommunications and Sensing,* ISBN-13: 978-0890063446, Boston: Artech House, June 1999

Qi, B.; Pickrell, G. R.; Xu, J.; Zhang, P.; Duan, Y.; Peng, W.; Huang, Z.; Huo, W.; Xiao, H.; May, G. R. & Wang, A. (2003). Novel data processing techniques for dispersive white light interferometer, *Opt. Engrg.,* Vol.42, No.11, (April, 2003), pp. 3165-3171.

Schuler, S.; Habel, W. & Hillemeier, B. (2008). Embedded fibre optic micro strain sensors for assessment of shrinkage at very early ages, *International Conference on Microdurability 2008,* pp.1377-1387, ISBN: 978-2-35158-065-3, Nanjing, China, 13-15 October 2008

Taylor, H. F. (2008). *Fibre Optic Sensor (second edition),* pp. 35-64, CRC Press, ISBN: 1420053655, Boca Raton, 2008

Thomson, W. (1857). On the electro-dynamic qualities of metals, *Proc. R. Soc.,* Vol. 8, pp. 546-50.

Udd, E. (1991). *An Introduction for Scientists and Engineers, Fibre Optic Sensors,* New York: John Wiley and Sons, 1991

Vengsarkar, A. M.; Perdrazzani, J. R.; Judkins, J. B.; Lemaire, P. J.; Bergano, N. S. & Davidson, C. R. (1996). Long-period fibre gratings based gain equalizers, *Opt. Lett.,* Vol. 21, No. 5, (March 1996), pp. 336-338.

Wnuk, V. P.; Méndez, A.; Ferguson, S. & Graver, T. (2005). Process for mounting and packaging of fibre Bragg grating strain sensors for use in harsh environment applications, Proc. of SPIE, Vol. 5758, (March 2005), pp. 46-54.

Xiao, H.; Zhao, W.; Lockhard, R. & Wang, A. (1997). Absolute Sapphire optical fibre sensor for high temperature applications, *Proc. SPIE.,* Vol. 3201, (April 1997), pp. 36-42.

Zhang, B. & Kahrizi, M. (2007). High-temperature resistance fibre Bragg grating temperature sensor fabrication, *IEEE Sensors Journal*, (April, 2007), Vol. 7, No.4, pp. 586 – 591.

Zhang, H.; Tao, X. M.; Yu, T. X.; Wang, S. Y. & Cheng, X. Y. (2006). A novel sensate 'string' for large-strain measurement at high temperature, *Meas. Sci. Technol.*, (February 2006) Vol. 17, No. 2, pp.450-458.

8

Bond-Based Earthquake-Proof of RC Bridge Columns Reinforced with Steel Rebars and SFCBs

Mohamed F.M. Fahmy[1] and Zhishen Wu[2]
[1]Assiut University
[2]Ibaraki University,
[1]Egypt
[2]Japan

1. Introduction

Bond between reinforcement and the adjoining concrete has been extensively studied, and it is confirmed that the use of deformed bars is essential for composite behavior of reinforced concrete (RC) structures. But since bond between the longitudinal bars and concrete results in concentration of damage at a specific localized interval of longitudinal bars where the local buckling occurs, Takiguchi et al. (1976) suggested mitigating this concentration of damage through unbonding of the longitudinal bars from concrete at plastic hinge zone. Kawashima et al. (2001) conducted an experimental study on RC columns reinforced with different lengths of unbonded bars at the plastic hinge zone. It was noticed that the failure of concrete was much less in the unbonded column than standard column, and strain on unbonded bar was less than that on the reinforcement of standard column. Recently, to improve the seismic performance of RC members, it is highlighted in the study of Pandey and Mutsuyoshi (2005) that reducing bond strength between the longitudinal bars and concrete has a favored effect on the failure mode, shear capacity and ductility of RC bridge piers: failure mode at ultimate state is changed from shear to flexural and shear strength and ductility are increased.

In the performance-based design approach, the design is primarily focused on meeting a performance objective, which is in line with a desired level of service (Floren et al., 2001, Priestley et al., 2007). For instance, new seismic design philosophies for bridges recommend that important bridges subject to massive earthquakes should be able to sustain the expected maximum lateral force in the inelastic stage with limited damages. To achieve this aim, structure should realize the existence of post-yield stiffness, damage level should be limited, and its permanent deformations (residual deformations) should be smaller than a specified limit; and all these indices are essentially dependent on the composite behavior of RC structures. On the other hand, the studies of Kawashima et al. (2001) & Pandey and Mutsuyoshi (2005) revealed the importance of reducing concrete-to-steel bond to mitigate the concentrated damage in the plastic hinge zone.

In the last two decades, civil engineers and designers have attempted to develop and adopt new forms of materials that would assist in the building of stronger, larger, more longer lasting, and aesthetic structures.Because the advantages of advanced composite materials, i.e., fiber reinforced polymers (FRPs), include: light weight, high strength or stiffness-to-weight ratios, corrosion resistance, and, in particular, the elastic performance, steel bars hybridized in the longitudinal direction with FRPs were recommended in the study of Fahmy et al. (2010) as an innovative reinforcement method for recoverable structures, where strain-hardening behavior of the innovative bars can be controlled based on both the amount and type of fibers used. Also, the use of the innovative rebars, i.e., steel fiber composite bars (SFCBs), will increase the life spans of structures because the inner steel bar is protected against corrosion (Fig 1). Uniaxial and cyclic tensile behavior of the innovative rebars were experimentally tested by Wu et al. (2010).

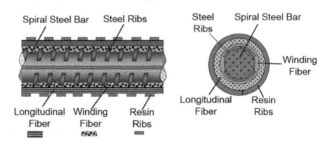

Fig. 1. Details of steel fiber composite bar (SFCB)

Due to the limited studies on bond-controlled structures which are reinforced with the ordinary steel rebars and because of the successful performance of the innovative SFCBs, it is essential to identify a suitable bond-based damage-controllable system, which guarantees limitation of the damage and mitigation of the permanent deformations. In addition, application of this system should not affect the structure load carrying capacity in the inelastic stage: structure should continue able to carrying load in the inelastic stage to withstand strong earthquakes. Hence, in the following, analytical studies using a computer program (Open System for Earthquake Engineering Simulation (Open SEES) [Mazzoni et al.]) are conducted, where effects of concrete-to-steel bond properties on the performance of RC bridge columns reinforced with rebars having different strain-hardening levels are determined. Effects of different bond conditions on column elastic and post-yield stiffnesses, residual deformations, and damaged zone are addressed. Validity of the analytical findings is established based on the experimental results of columns reinforced with unbonded deformed steel bars (DSBs) in the plastic hinge zone by Kawashima et al. (2001), columns with bond-controlled reinforcements by Pandey and Mutsuyoshi (2005), and two columns reinforced with rebars having a controlled strain-hardening behavior, i.e. steel fiber composite bars. Ultimately, the influence of concrete-to-SFCBs bond on the recoverability of RC bridge columns is analytically studied.

2. Idealized load-deformation model of damage-controllable RC structures

Fig. 2 shows a mechanical model of damage-controllable RC structures located in high seismicity zones. The proposed model exhibits the required performance from newly constructed structures under the effect of different levels of seismic load, where the lateral response proceeds along O-A-B-C-D-E-F-G. The behavior of a general RC flexural structure whose lateral response is along O-A'-B'-C'-D'-F' is also given for comparison. Prior to the yielding of steel reinforcement, lines OAB and OA'B' corresponding to both types of structures share similar stiffnesses, K_1. The most remarkable difference occurs after the yielding of the steel reinforcement: after point C and C'. For the general RC structure, the deformation increases dramatically almost without any significant increase in load carrying capability: along line C'D' almost zero post-yield stiffness is demonstrated. However, with the proposed approach, the structure can still carry the load even after the steel reinforcement yields and hardening behavior has been exhibited along line CD. Based on the codes requirements for ductile structures to withstand strong earthquakes, the proposed structure is characterized by the part DEF after the hardening zone, where favorable ductility is demonstrated. The ultimate drift (δ_u) corresponding to point F or F' for the proposed structure and the general RC structure, respectively, is defined for both structures to be at 20 percent strength decay, (Park & Paulay, 1975).

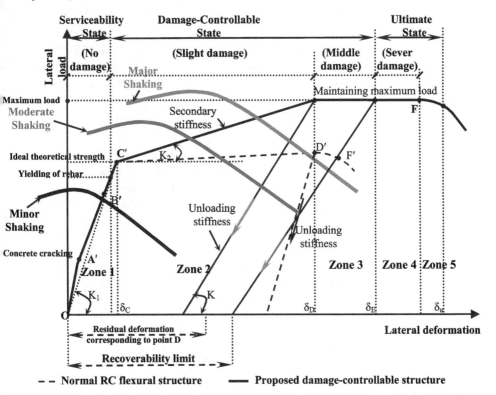

Fig. 2. Idealized load-deformation behavior of proposed damage-controlled structures

According to the mechanical behavior shown in Fig.2, the load-deformation of the proposed structure can be divided into five main zones; Zone 1: from point O to B; Zone 2: from point B to D; Zone 3: from point D to E; Zone 4: from point E to point F; and Zone 5: after point F. Zone 1 corresponds to a stage of no damage or concrete cracking. Under a small earthquake, the mechanical behavior should be controlled in this zone, and the original function of the structure can be maintained without any repair and displacement of elements. Zone 2 corresponds to the hardening behavior after the yielding of steel reinforcements, where a distinct secondary stiffness is demonstrated and the dramatic deformation can be effectively controlled. Under a medium or strong earthquake, the mechanical behavior of the proposed structure should be within zone 2. Thus, damage can be effectively controlled by the secondary stiffness. The original function of the structures can be quickly recovered through repairs after a medium or large earthquake. Zone 3 corresponds to ductile behavior after hardening, where favorable ductility is demonstrated under a large earthquake but with middle level of damage. Furthermore, within this zone structure should be recoverable, where residual deformation does not exceed recoverability limit, i.e. residual deformations should be less than 1% of structure height for quick recovery of original functions of structure after an earthquake. Zone 4 corresponds to additional ductile behavior, where the proposed structure can be kept in place without collapse during a large earthquake, though severe damage may occur. The original function of the structures may be recovered through the replacement of some elements. During a severe earthquake, the mechanical behavior may enter zone 5 with collapse.

It is clear that two indices should be applied to measure the recoverability of RC structures: secondary stiffness and residual deformation. However, to have a complete description of structure performance, damage level should be considered in the evaluation of the recoverability of structures after seismic excitation. In the study of Mostafa (2011) several damage measures based on a single response parameter are summarized.

3. Analytical investigation

Of crucial importance in the anti-seismic design of RC bridges is to ensure the gradual increase of strength after yielding (existence of positive post-yield stiffness) and to minimize the permanent deformation due to a massive earthquake so that the damage can be easily repaired. That is quickly recoverable bridges after sever seismic actions. In this study mechanical properties of steel reinforcement used and its bond conditions to the surrounding concrete are key parameters to investigate how could be the required recoverability of RC bridges controlled? Thus, in the following section, analytical studies are carried out to detect their effects on the performance of RC columns: elastic and post-yield stiffnesses, residual deformations, and damage level.

3.1 Fiber-based model of RC columns with zero-length section element

Cyclic loading analysis is conducted using a computer program (Open SEES) [Mazzoni et al.]. This program has a variety of predefined material models for multiple applications that can be manipulated to fit specific criteria and properties. Since the fiber analysis remains the most economic and accurate means to capture seismic behavior of concrete structures (Spacone et al. (1996) and Saiidi et al. (2009)), the fiber model was used, Fig. 3. The confined concrete properties were based on Kent-Scott-Park's model. The widely used Giuffré-Menegotto-Pinto model is employed in this study to represent the hysteretic stress-strain behavior of longitudinal steel reinforcement. The model includes the yielding, strain hardening, and Bauschinger effect of the steel bar.

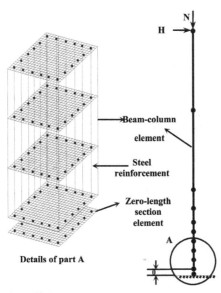

Fig. 3. Fiber-based modeling of the studied columns

Zhao & Sritharan (2007) have developed fiber-based analysis of concrete structure through the incorporation of a zero-length section element to reflect the effect of the fixed-end rotation that arises at the column-foundation or column-beam interface on the performance of structures. The conducted analyses by Zhao & Sritharan (2007) on cyclic responses of cantilever columns and a bridge tee-joint system, satisfactorily captured deflections, force versus displacement hysteresis responses, and strains in the longitudinal reinforcing bars. The developed constitutive model by Zhao & Sritharan (2007) for the steel fibers of the zero-length section element expresses the bar stresses (σ) versus loaded-end slip (S) response. The main parameters of this model are the loaded-end slips S_y and S_u and their corresponding bar stresses σ_y and σ_u, respectively. Incase the bar has a sufficient anchorage length, f_y and f_u are the stresses corresponding to S_y and S_u, respectively, where f_y and f_u are the yield and ultimate strengths of the steel reinforcing bar, respectively, Fig. 4(a). Zhao & Sritharan (2007) defined that the sufficient anchorage length is not less than $((d_b/7)(f_y/(f'_{co})^{0.5})$, where d_b is the bar diameter (mm) and f'_{co} is the concrete compressive strength in MPa. To determine the suitable value of S_y, Zhao & Sritharan (2007) proposed the following experimentally-based equation:

$$S_y = 2.54 \left[\frac{d_b}{8437} \frac{f_y}{\sqrt{f'_{co}}} (2\alpha + 1) \right]^{1/\alpha} + 0.34 \tag{1}$$

where a was taken as 0.4. Definition of S_u value is based on the determined value of S_y: $S_u = 30 \sim 40\ S_y$. This model is employed for capturing the slip effect in flexural members subjected to reversed cyclic loading, and hysteretic rules were established, Fig. 4(b). Consequently, this model is adopted here to find out the impact of different bond behaviors on the performance of cyclically loaded RC columns.

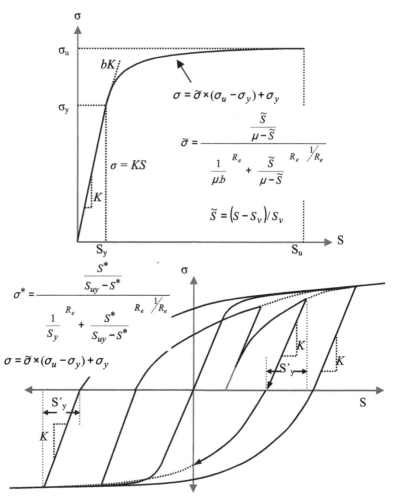

Fig. 4. (a) Envelop curve and (b) Hysteretic model for steel bar stress versus loaded-end slip relationship (Zhao & Sritharan, 2007).

3.2 Effect of bond parameters and rebar mechanical properties on the performance of RC column

Kawashima et al. (2001) conducted an experimental study on four RC columns (A, B, C, and D). Specimen A was tested to simulate the performance of conventionally reinforced column but the others were reinforced with bars unbonded at the plastic hinge zone. Here, hysteretic response of Specimen A is considered as a control sample to find out the effects of different bond conditions between column reinforcement and the surrounding concrete on its hysteretic performance (in particular the required recoverability). The tested specimen had square cross-section of 400x400mm and 1450-mm-height. It was reinforced with twelve longitudinal rebars of 13-mm-diameter of 367 MPa yield strength, lateral reinforcement consisted of 6-mm-diameter ordinary rebars with yield strength of 376 MPa and spacing of 50mm. The tested compressive strength of the concrete cylinder was 20.6 MPa. The column was tested under constant axial load (160kN) and reversed cyclic lateral load. Using the four parameters of Zhao and Sritharan model ((S_y and σ_y) and (S_u and σ_u)), ten different bond conditions are studied (Table 1).

Group	Case	Column	Steel mechanical properties			Bond properties			
			f_y (MPa)	f_u (MPa)	E_2/E_1	S_y (mm)	σ_y (MPa)	S_u (mm)	σ_u (MPa)
I	1	$0.01\text{-}S_y\text{-}0.8f_y$	367	$1.5\,f_y$	0.01	S_y	f_y	S_u	$0.8f_y$
	2	$0.01\text{-}2S_y\text{-}0.8f_y$	367	$1.5\,f_y$	0.01	$2S_y$	f_y	$2S_u$	$0.8f_y$
	3	$0.01\text{-}3S_y\text{-}0.8f_y$	367	$1.5\,f_y$	0.01	$3S_y$	f_y	$3S_u$	$0.8f_y$
II	4	$0.01\text{-}S_y\text{-}f_y$	367	$1.5\,f_y$	0.01	S_y	f_y	S_u	f_y
	5	$0.01\text{-}2S_y\text{-}f_y$	367	$1.5\,f_y$	0.01	$2S_y$	f_y	$2S_u$	f_y
III	6	$0.01\text{-}S_y\text{-}1.5f_y$	367	$1.5\,f_y$	0.01	S_y	f_y	S_u	$1.5\,f_y$
	7	$0.02\text{-}S_y\text{-}2f_y$	367	$2.0\,f_y$	0.02	S_y	f_y	S_u	$2f_y$
IV	8	$0.04\text{-}S_y\text{-}2f_y$	367	$2.0\,f_y$	0.04	S_y	f_y	S_u	$2f_y$
	9	$0.08\text{-}S_y\text{-}2f_y$	367	$2.0\,f_y$	0.08	S_y	f_y	S_u	$2f_y$
V	10	$0.12\text{-}3S_y\text{-}1.5f_y$	367	$2.5\,f_y$	0.12	$3S_y$	f_y	$3S_u$	$1.5f_y$

Note: E_1= 200 GPa

Table 1. Reinforcement mechanical properties and corresponding bond properties

To evaluate the effect of the strain-hardening behavior of steel reinforcement used on the performance of bond-controlled structures, five values of the bilinear ratio factor ($r = E_2/E_1$), where E_2 and E_1 are the steel elastic and post-yielding stiffnesses, are examined here, i.e. $r = 0.01$, 0.02, 0.04, 0.08, and 0.12. Of course bond between reinforcement and the adjoining concrete is an essential factor in the definition of both σ_y and σ_u and the corresponding slip values. Therefore, five groups (I, II, III, IV, and V) are studied here, Table 1. The groups I, II &III consider the effects of bond parameters on the performance of columns reinforced with ordinary rebars (bilinear ratio factor = 0.01). In case no. 6 of group III, bond parameters are those proposed by Zhao & Sritharan (2007) to include the effect of strain penetration of longitudinal bars into foundation on the lateral response of conventionally RC columns. Meanwhile, the aim in the groups I and II is to characterize the effect of weak bond strength between ordinary rebars and concrete, so, it is assumed that rebars may not approach the ultimate strength, e.g. $\sigma_u \leq f_y$, and also S_y would increase to two or three times of its value evaluated based on Eq. 1. Effect of higher bilinear ratios of steel reinforcement is examined in the last two groups (IV and V). In group IV, the ultimate strength of rebars used is kept constant ($2f_y$) while different bilinear ratios ($r = 0.02$, 0.04 and 0.08) are studied. In the last case study (group (V)), column is reinforced with rebars having bilinear ratio factor of 0.12 but it is assumed that, through controlling the bond conditions, rebars may not approach more than 60% of their ultimate strength along with the effect of additional slippage. It is noteworthy that the studied cases of RC columns are typified based on the bilinear ratio between post-yield and elastic stiffnesses of steel reinforcing bar, rebar loaded-end slip at σ_y, and maximum achieved strength of reinforcing steel at the ultimate slip (S_u) of rebars. For instance, ($0.01\text{-}2S_y\text{-}0.8f_y$) is the designation of case two of group I, where 0.01 is the bilinear ratio factor of reinforcing steel; $2S_y$ is double of rebar slip value calculated by Eq. 1 and σ_y is the corresponding strength; and $0.8f_y$ is 80% of rebar yield strength, which is considered the maximum achieved strength (σ_u).

Hysteretic responses of the studied cases are depicted in Fig. 5, wherein the measured force-drift relationship of Specimen A is superimposed for comparison.

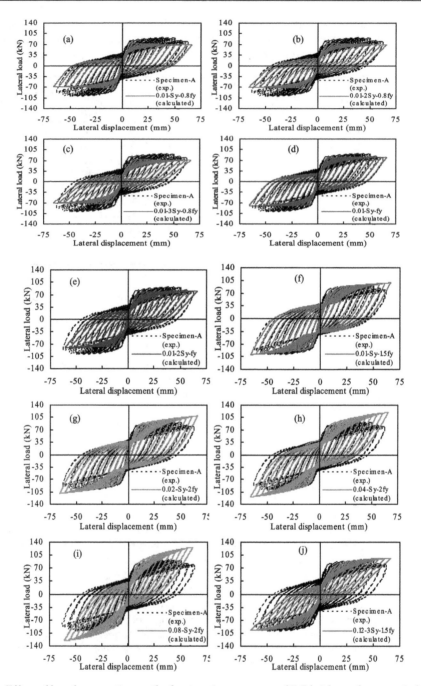

Fig. 5. Effect of bond properties on the hysteretic responses of RC bridge columns reinforced with rebars of different levels of strain-hardening

5.2.1 Detailed discussion on bond-based performance of RC columns

In the light of the aforementioned analytical results, this section presents a detailed discussion about the effects of altering bond conditions and steel type on the performance of RC column in terms of its post-yield stiffness, residual deformation, and damage level.

5.2.1.1 Post-yield stiffness

The post-yield stiffness is defined as the gradual increase of column strength after fulfilling its theoretical moment capacity M_i, and the end point of the post-yield stiffness is that corresponding to the maximum lateral capacity. The theoretical moment capacity is calculated using ACI rectangular stress block for concrete in compression, which has a mean stress of $0.85f'_{co}$, the measured concrete compression strength f'_{co} and steel yield strength f_y, and an ultimate concrete compression strain of 0.003 [ACI 318-08]. Based on the specified details, dimensions, and material properties of the tested specimen A, theoretical ideal strength ($P_i = M_i/L$, where L is column height) is defined and superimposed in Figs. 6 (a-d). Figs. 6 (a-d) show the skeleton curves of the hysteretic responses of the analytical studied cases.

The experimental results of the conventionally tested column (Specimen A) showed that column was able to carry the load even after achieving the theoretical strength (84.4 kN) and hardening behavior has been exhibited. The envelope of the measured force-drift relationship of Specimen A is superimposed on the results of the studied cases, as shown in Figs. 6 (a-d). The curves are the average of the envelopes for the push-and-pull loadings. It was reported by Pettinga et al. (2006) that systems exhibiting post-yield stiffness ratios greater than 5% will have significantly reduced permanent displacements. The ratio between column post-yield stiffness (k_2) and elastic stiffness (k_1) of Specimen A is 2.35%. Accordingly, to achieve the aim of quickly recoverable bridges after massive earthquakes, an enhancement of the performance of conventionally RC column is still necessary.

When weak bond strength between ordinary rebars ($r = 0.01$) and the surrounding concrete is considered, the analytical results show that the column performance in the inelastic stage is function of reinforcement stresses at the ultimate slippage, Figs. 6 (a & b). For instance, in the studied cases of group I, where reinforcement stresses would not exceed 80% of the yield strength, column deformation increases dramatically with a decrease in the load carrying capability after yielding (negative post-yield stiffness), Fig. 6 (a); but, In group II, zero post-yield stiffness is the column performance in the inelastic stage when reinforcement stresses would approach the yield strength at the ultimate slippage, Fig. 6 (b). It is noteworthy that, in both groups, columns could not reach up to the theoretical strength of Specimen A.

As is clear from Fig. 6 (c), replacement of ordinary rebars with others, having higher strain-hardening levels and perfect bond to the surrounding concrete, would enhance column performance in the inelastic stage. Where the increase in r value leads to an increase in column post-yield stiffness, e.g. the ratios between column post-yield and elastic stiffnesses (k_2/k_1) are 2.88%, 4.05%, and 5.2% when steel bilinear ratio are 0.02, 0.04, and 0.08, respectively. It should be noted that the first stiffness was the same among the studied cases and Specimen A, as shown in Fig. 6 (c).

In the last case study, where r is 0.12, although merely 60% of rebar ultimate strength could be achieved along with an increase in rebars slippage, Fig. 6 (d) shows that columns could realize the existence of post-yield stiffness ($k_2/k_1 = 3.18\%$) and reach up to the same lateral strength of specimen A. Moreover, in a distinction from the other studied cases, a third zone with zero post-yield stiffness characterizes the inelastic stage of this column.

It is noteworthy that the slope of column elastic stiffness would slightly affect by the increase in rebar slippage at yielding when it is double of the value calculated from Eq. 1, Figs. 6 (a & b); however, it decreases to two-third of the elastic stiffness of conventionally RC column when rebar slippage increases to $3S_y$, regardless of whether the bilinear ratio is small or high, Figs. 6 (a & d). The lower stiffness of the columns would lead to longer vibration period for the columns and would generally reduce earthquake forces (Saiidi et al. (2009)).

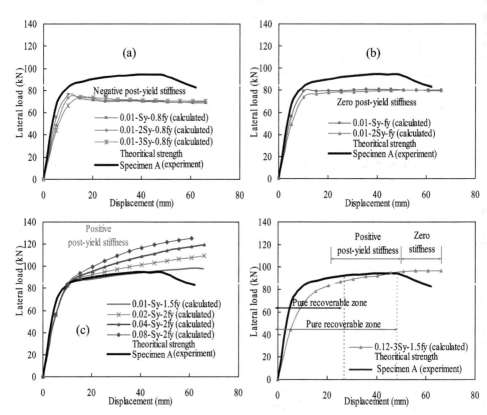

Fig. 6. Effect of bond properties on the post-yield stiffness of RC bridge columns reinforced with rebars of different levels of strain-hardening

3.2.1.2 Residual deformations

Using the test results of specimen A and analytical results of the studied cases (Fig. 5), the drift ratios versus columns residual drift ratios are plotted in Fig. 7. The residual deformation, which is defined as the displacement of zero-crossing at unloading on the hysteresis loops, should not exceed 1% of the column height for a quick recovery after an earthquake (JSCE, (2000)). For specimen A, the drift ratios versus residual drift ratios are superimposed on Figs. 7 (a–d) for comparison. While specimen A was able to reach up to 4.6% lateral drift ratio with almost 12.1% strength decay, the end of the recoverable stat is corresponding to 1.83% lateral drift, Fig 5 (d) & Fig. 7. After lateral column drift of 1.83%, regardless of whether damage level is light, medium, or sever, column can not continue to function following major earthquakes and demolishing may be required. Consequently, mitigation of column residual deformations is critical to have ductile-recoverable bridges.

Figs. 7 (a & b) present the residual drift ratios verses lateral drift ratios of columns with different levels of bond capacity between ordinary rebars and the surrounding concrete. As is clear, mitigation of column residual deformation is in direct relationship with rebar slippage, where the increase in rebar slippage to two and three times of the value defined by Zhao & Sritharan model shifts the recoverable zone by almost 0.5% and 1% lateral drift, respectively.

The effects of using reinforcement with high bilinear ratio (r) on column residual deformations are shown in Fig. 7 (c), where a favorable effect could be noticed. For example, 2.16% lateral drift is the end of the recoverable state of column reinforced with rebars of 0.02 bilinear ratio factor; and when the bilinear ratio increases to 0.04 and 0.08, the end of the recoverable stat is shifted to 2.38% and 2.67%, respectively. Furthermore, Fig. 7 (d) exhibits the effect of using rebars with a higher bilinear ratio ($r = 0.12$) coupled with the effects of weak bond strength to the adjoining concrete. The simulation results indicate that column performance is ductile-recoverable, where column could be recoverable until lateral drift ratio of 3.34% with the existence of post-yield stiffness, as shown in Fig. 6 (d).

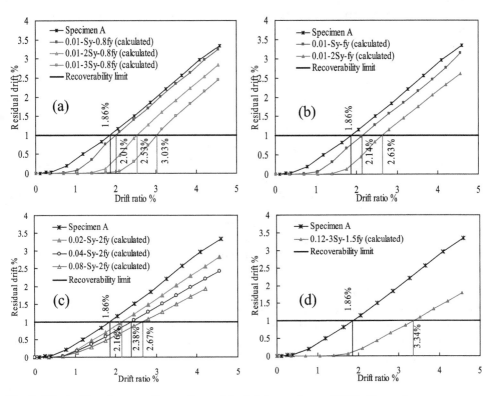

Fig. 7. Effect of bond properties on the residual deformations of RC bridge columns reinforced with rebars of different levels of strain-hardening

3.2.1.3 Damage level

The main conclusion taken from the aforementioned analytical results presented in Figs. 6 and 7 is that both steel strain-hardening behavior and concrete-to-rebar bond capacity are key parameters, which can be employed to achieve the required performance from structures during and after seismic actions. Even so, to define a suitable system guaranteeing the aim of damage-free recoverable structures, investigating the effects of the studied parameters (Table 1) on column damage level is indispensable. The literature on damage measures of structures under ground motions is vast (e.g. Cosenza et al. 1993; Ghobarah et al. 1999; Abbas 2011). Damage indices are based on either a single or combination of structural response parameters.

To have a structure capable of deforming in a ductile manner when subjected to several cycles of lateral loading, deformations concentrate in the region of the primary lateral force resisting mechanism (plastic deformation zone), wherein the demand of high deformability (curvature capacity) increases with the increase in the ductility demand. Consequently, length of this zone, which is one of the crucial aspects in the performance-based design, along with the distribution of curvature ductility (φ/φ_y) would be useful to find out the impact of the studied parameters on the damage level, because the curvature (φ) is dependent on steel strains. φ is calculated using $(\varepsilon_{st}-\varepsilon_{sc})/(d-d')$, where ε_{st} and ε_{sc} are the steel strains in the tension and compression sides of the loaded column, respectively (tensile strains are taken as positive and compression strains as negative), $d-d'$ is the distance between the tension and compression steel pieces, and φ_y is the curvature at first yield.

3.2.1.3.1 Depth of yielding zone (Plastic deformation zone)

At the ultimate achieved lateral drift of Specimen A, which is almost 4.2%, length of the plastic deformation zone is defined for each of the studied cases in Table 1 based on the calculated distribution of curvature ductility through the column height. Fig. 8 shows the depth of the yielding zone as a ratio of the column height for all the studied cases. It is clear from the figure that depth of the yielding zone is in direct proportion to the value of r of reinforcement used when it is in perfect bond to the surrounding concrete, where the normalized depth of the yielding zone could be changed from 0.41 to 0.54 by the increase of r from 0.02 to 0.08. In addition, it is evident that depth of the yielding zone is significantly affected by bond conditions, where the smallest depth of the yielding zone would be for columns with weak bond strength between ordinary rebars and the adjoining concrete, and this depth may possibly not change by the increase in the slippage of rebars.

3.2.1.3.2 Curvature distribution

While the yielding zone may locate in a small region at column base or extend beyond the normal range of the yielding zone based on the reinforcement type and its bond condition to the surrounding concrete, the worst case, i.e. with high damage level, still can not be defined. For example, the column typified $(0.08-S_y-2f_y)$ has the greatest normalized depth of the yielding zone ($0.54 > 0.36$ of conventionally reinforced column) among the studied cases, however damage level would not be the worst based on the curvature distribution of both columns. Hence, distribution of curvature ductility is determined for all of the studied cases.

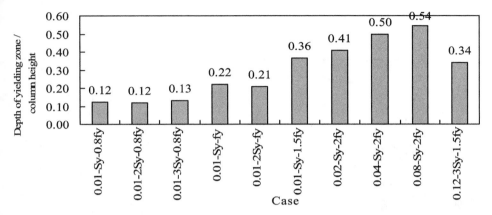

Fig. 8. Depth of the yielding zone of the studied cases at drift of 4.2% of column height

When bond strength is sufficient to assure the composite behavior between concrete and rebars, steel greatly contributes to the required deformability in the plastic zone. This is clear from Fig. 9 (a), where conventionally reinforced columns (case no. 6) exhibits high concentration of curvature ductility within the first 250 mm of the column height, i.e. curvature ductility is 32.3 at column base and reduces to 5.1 at 250mm. On the other hand, steel contribution to the required curvature ductility reduces when concrete-to-steel bond is weak. This can be seen from Fig. 9 (b) where curvature ductility reduces to 14.6% and 4% of that achieved by the conventionally reinforced column at its base when steel stresses at the ultimate slippage are f_y and $0.8f_y$, respectively, regardless of rebar ultimate slippage value. However, those columns can achieve the same level of the lateral drift of Specimen A as shown in Figs. 5 & 6, and this could be attributed to the increase in the contribution of column fixed-end rotation to lateral column deformation as shown in Fig. 10 (cases 1 to 6, listed in Table 1). In other words, when bond strength is weak, the reduction in steel contribution to lateral column deformation is compensated through the contribution of the fixed-end rotation due to rebars slippage, and this in turn reveals mitigation of damage in the plastic deformation zone.

Fig. 11 (a) details the distribution of curvature ductility of columns reinforced with rebars having bilinear ratio factors higher than that of ordinary rebars, e.g. rebars with $r = 0.02$, 0.04, and 0.08, where perfect bond with the adjoining concrete are assumed for the three cases. It is clear that curvature ductility at column base and up until a height of 250 mm reduces by the increase in the value of r; nevertheless, out of this region the increase in r adversely affects the distribution of curvature ductility: at 400mm of column height, the change in r from 0.02 to 0.04 and 0.08 increases curvature ductility from 1.8 to 4.1 and 3.8, respectively, and at 600 mm it increases from 0.97 to 1.16 and 1.3, respectively. Furthermore, in these three cases the curvature ductility demands within 100-mm from the column base are smaller than those of the conventionally reinforced column (Fig. 9 (a)). In conclusion, using rebars with high level of strain-hardening may contribute to the mitigation of damage in the plastic hinge zone as a result of the redistribution of curvature ductility through the column height, and this propagation of yielding to a higher depth would compensate the demand of high deformability at column base to achieve a certain level of lateral drift.

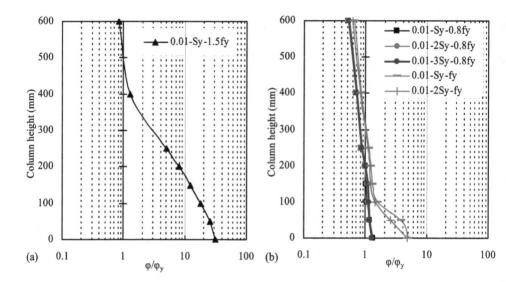

Fig. 9. The effect of bond properties between ordinary rebars ($r = 0.01$) and concrete on distribution of curvature ductility through column height at lateral drift of 4.2%

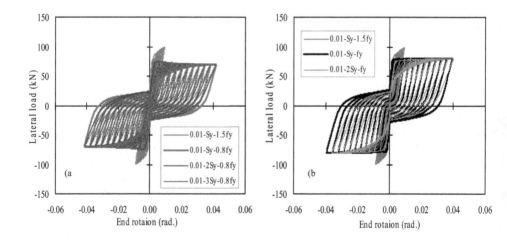

Fig. 10. Calculated fixed end-rotation versus lateral load

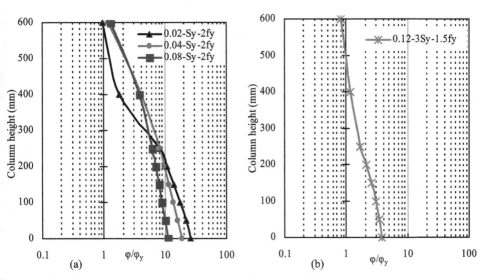

Fig. 11. The effect of bond properties between rebars ($r > 0.01$) and concrete on distribution of curvature ductility through column height at lateral drift of 4.2%

The combined effect of both high bilinear ratio and bond capacity on the distribution of curvature ductility can be evaluated from Fig. 11 (b), where a well distribution of column curvature could be achieved. Curvature ductility is almost 3.78 at the interface section, is reduced to 2.12 and 1.17 at 200 mm and 400 mm from the column base, respectively, and ends at 0.8 at 600 mm. Curvature distribution in this case also reveals the importance of using rebars with high strain-hardening performance, because when ordinary rebars loss their abilities of carrying load after yielding (cases number 4 & 5 in Table 1) due to the weak bond strength, concentration of damage very close to column base is highly probable, see Fig. 9 (b).

Based on the aforementioned analytical results, contribution of steel deformations to lateral deformation of RC columns should be reduced to mitigate damage at the zone of plastic deformations and thus contribution of the fixed-end rotation, due to rebars slippage, is critical to achieve the desired drift demand. For conventionally reinforced structures, contribution of column fixed-end rotation could only guarantee mitigation of the concentrated damage in plastic deformation zones, namely, structures reinforced with rebars having elastic perfectly-plastic behavior or small bilinear factor ($r \leq 0.01$), this contribution would be accompanied by a negative or zero post-yield stiffness behavior, Fig.6 (a&b). However, reinforcing structures with bond-controlled rebars having high bilinear ratio factor would be a reasonable system to have a damage-free recoverable structure. Because structure might be able to continue carrying load in the inelastic stage and both damage level and residual deformations could be controlled.

4. Experimental evaluation of recoverability of RC bridge columns reinforced with bond-controlled rebars and SFCBs

In order to verify the analytical findings, experimental results of columns reinforced with unbonded deformed bars in plastic hinge zone (Kawashima et al., 2001) and those with bond-controlled reinforcements (Pandey & Mutsuyoshi, 2005) are restudied here from the view point of the required recoverability. Furthermore, two columns reinforced with SBFCBs and SCFCBs were tested by Fahmy et al. (2010) to examine the effect of high strain-hardening behavior of reinforcement used on the required recoverability. Reinforcements of both specimens were in normal bond conditions to the surrounding concrete.

4.1 Bridge columns reinforced with unbonded deformed bars

To enhance ductility of RC bridge columns, Kawashima et al. (2001) conducted an experimental study on scale-model RC columns reinforced with ordinary deformed bars, which were unbonded at the plastic hinge zone. Different unbonding lengths were considered in their study, e.g. 200mm, 400mm, and 600mm were the unbonded lengths in Specimens B, C, and D, (Fig. 12, and Table 2), respectively. The tested specimens had square cross-section of 400x400mm and 1450-mm-height. In view of the results of Kawashima et al. (2001), the required recoverability of unbonded RC columns are examined in this study.

The envelopes of the measured force-drift relationships of the three unbonded specimens are shown in Fig. 13. The curves are the average of the envelopes for the push-and-pull loadings. Additionally, for comparison, the measured force-drift relationship of Specimen A is superimposed on the same figure. It is evident from Fig. 13 that the increase of the unbonded length is accompanied by a decrease in the elastic stiffness, reduction of the achieved ultimate strength, and diminishing of post-yield stiffness.

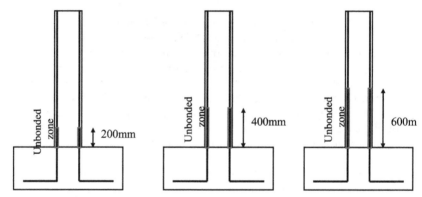

Fig. 12. Schematic details of the unbonded length of RC bridge columns

Author	Sample	Unbonded length (mm)	Axial Load, ratio (%)[a]	Main steel	f'_{co} (MPa)	Yield stress of main steel, (MPa)	Yield stress of lateral steel, (MPa)
Kawashima et al. (2001)	Specimen B	200	4.2	12 D13	23.6	367	376
	Specimen C	400	4.1	12 D13	24.6	367	376
	Specimen D	600	4.2	12 D13	23.5	367	376

[a] axial load ratio = ((axial load)/($A_g . f'_{co}$)), where A_g is the gross section area.
D13 = deformed bar with diameter of 13 mm

Table 2. Details, dimensions, and material properties of unbonded RC columns

Residual drift ratios verses column lateral drift ratios are plotted in Fig. 14 for the three unbonded specimens and the control one. This figure shows that there is a close similarity in the relation between the column drifts and the corresponding residual deformations for the four columns, which reveals that unbonding of deformed rebars at the plastic hinge zone with any length could not assure the alleviation of column residual deformations.

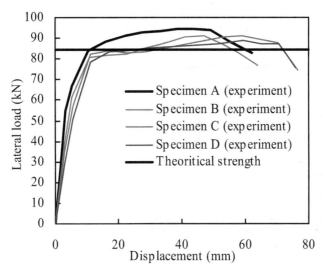

Fig. 13. Skeleton curves of the hysteretic responses of columns reinforced with bonded and unbonded deformed bars

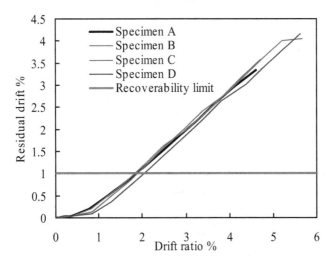

Fig. 14. Residual deformations of columns reinforced with bonded and unbonded deformed bars

4.2 Bridge columns reinforced with bond-controlled rebars

An experimental investigation was carried out by Pandey and Mutsuyoshi (2005) to examine how controlling the bond of the longitudinal reinforcements can improve seismic performance factors, such as shear strength and ductility, of RC structures. Results of three of the tested columns (C-1, C-6, and C-7) are examined here to find the effects of surface characteristics of rebars used and also bond conditions of steel-to-concrete on the required recoverability: columns C-1 and C-6 were reinforced with DSBs and RSBs (round steel bars), respectively, but column C-7 was reinforced with RSBs, which had been coated with grease to further reduce bond strength through 800 mm length. Details of those columns are given in Table 3.

Author	Sample	Column Height, (mm)	Unbonded length (mm)	Axial Load, ratio (%)[a]	Main steel	Transverse steel (Size and spacing, mm)	f'_{co} (MPa)	Yield stress of main steel, (MPa)	Yield stress of lateral steel, (MPa)
Pandey and Mutsuyoshi (2005)	C-1	1050	----	2.7	12 D16	D6 at 150	36.4	396.4	426.7
	C-6	1050	----	2.6	12 Φ16	D6 at 150	39.21	298.6	426.7
	C-7	1050	800	2.4	12 Φ16	D6 at 150	42.52	298.6	426.7

[a] axial load ratio = $((\text{axial load })/(A_g \cdot f'_{co}))$, where A_g is the gross section area.
D16 = deformed bar with diameter of 16 mm; and Φ16 = round bar with diameter of 16 mm.

Table 3. Details, dimensions, and material properties of bond-controlled RC columns

Because the lateral response of any of the investigated columns differs from that of others, the effect of the studied parameters on the achieved post-yield stiffness was evaluated as shown in Fig. 15, where average drifts of pull-and-push excursions of loading are plotted in relation to the corresponding column lateral strengths (P) divided by the theoretical strength (P_i). As a result of coating RSBs with grease, elastic stiffness of column C-7 was the smallest among the examined columns, Fig. 15. On the other hand, replacement of DSBs with RSBs slightly affected the elastic stiffness of column C-6, and it could not alter the shear failure mode of column C-1 to the ductile failure mode of column C-7, which was due to crushing and spalling of the concrete cover, followed by yielding of the longitudinal bars at the column-footing joint. In all cases shown in Fig. 15, negative post-yield stiffness is the performance in the inelastic stage, where the greatest deterioration of strength appeared in column C-1.

Relationship between column drifts and residual drifts is depicted in Fig. 16 for the columns C-1, C-6, and C-7. As is seen from Fig. 16, residual deformations of columns C-1 and C-6 are the same at any lateral drift, and their drifts at the recoverability limit are almost identical. This means that the use of RSBs in place of DSBs has no impact on column residual deformations. On the other hand, a favorable mitigation of column residual deformations was for column C-7 reinforced with unbonded RSBs. Column C-7 was capable of staying recoverable until lateral drift of 3.05%, which is almost 1.5 times the drift of column C-6 at the recoverability limit.

Although unbonding of DSBs at plastic hinge zone had no effect on column residual deformations, Fig. 14 and 16, it is interesting to stress on this finding which would help in realizing how could be the residual deformations controlled through the bond conditions between reinforcement and the surrounding concrete. When DSBs or RSBs are unbonded to the surrounding concrete, rigid body rotation due to rebar slippage would contribute to column lateral deformation and thus contribution of steel greatly reduces, and this was acknowledged in the study of Kawashima et al. (2001) through the measurements of steel strain at the plastic hinge zone, where strains of unbonded rebars were much lower than the counterpart in conventional RC columns. Despite the reduction of steel deformations, column residual deformations did not significantly change, particularly, when longitudinal reinforcement used was DSBs, and this can be attributed to partial unbonding of the DSBs, i.e. length of the unbonded zone, and characteristics of rebar surface. That is, owing to the interlocking of DSBs ribs of the bonded regions with the surrounding concrete, rebar cannot return to its former place after slippage. Therefore, unbonding of RSBs with grease coating for a length of ~ 0.76 of column height (column C-7) increased the opportunity of limiting the residual fixed-end rotation and in turn the permanent deformations. In the study of Pandey and Mutsuyoshi (2005), it was reported that unbonding of most column height is an appropriate technique to avoid initiation of flexural cracks from the bonded regions. Consequently, mitigation of residual deformations of bond-controlled RC columns is dependent on the geometry of rebar surface as well as the length of the bonded regions.

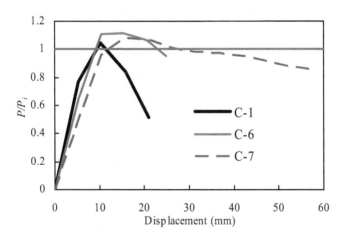

Fig. 15. Skeleton curves of the hysteretic responses of columns reinforced with bonded deformed bars and bonded and unbonded round bars

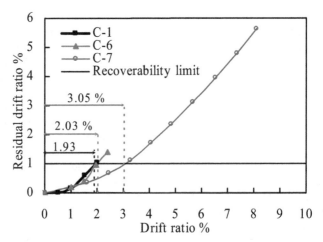

Fig. 16. Residual deformations of columns reinforced with bonded deformed bars and bonded and unbonded round bars

4.3 Bridge columns reinforced with steel fiber composite bars (SFCBs)

The results of this study are treated in a longer length in the study of Fahmy et al. (2010). Here, the hysteretic responses of the tested three columns are given to verify the possibility of enhancing recoverability of RC bridges using rebars with high level of strain-hardening as longitudinal reinforcement in place of ordinary rebars. Fig. 17 shows the results of three tested columns (CS14, CS10-C40 and CS10-B30), where the first column was reinforced with twelve deformed rebars with elastic perfectly-plastic behavior and the other columns were reinforced with steel fiber composite bars (r is over 10%). It is evident from Fig. 17 that conventionally reinforced column was able to achieve lateral drift until 30 mm without any significant lose in the lateral capacity. However, after ~21 mm later displacement (1.91% lateral drift ratio) the column fell in the irrecoverable stat. on the other hand, the results of the other two columns (CS10-C40 and CS10-B30) showed recoverable performance till lateral drifts of 3.06% and 3.16%, which correspond to 10.2% and 5.9% drop in the achieved ultimate strengths, respectively. However, due to the good bond between SFCBs and the adjoining concrete, Fig. 18 reveals extension of the plastic zone of column CS10-B30 beyond the counterpart in the conventionally reinforced column: depth of the damaged zone was almost 25% of the column height for the columns CS14 and 31% for CS10-B30, respectively.

Overall, the agreement found between all the addressed experimental results and the theoretical findings is acceptable. For instance, the decrease in the slope of both column elastic and post-yield stiffnesses due to poor bond qualities between ordinary rebars and the surrounding concrete is evident in the experimental results of Kawashima et al. (2001) and Pandey & Mutsuyoshi (2005), Figs. 13 and 15. Moreover, the enhancement in the inelastic stage with the existence of post-yield stiffness was successfully achieved by columns reinforced with SFCBs. The analytical results showed that column residual deformations could be reduced when rebars slippage at yielding increases or rebars used have high level of the strain-hardening, and both effects are validated by the results of column C-7 reinforced with RSBs coated with grease (Fig. 16) and the results of the columns CS10-C40 and CS10-B30 reinforced with SFCBs, respectively, see Fig. 17.

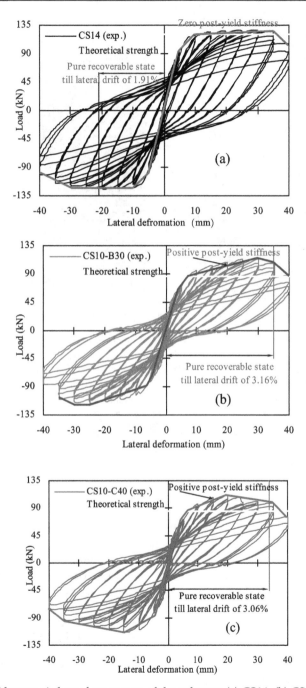

Fig. 17. Measured hysteretic lateral responses of the columns (a) CS14, (b) CS10-B30, and (c) CS10-C40

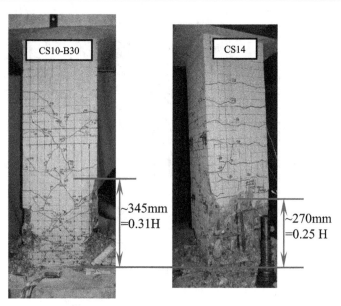

Fig. 18. Depth of damaged zone of the columns (a) CS10-B30 and (b) CS14

In conclusion, recoverable performance of RC columns lying in high seismicity regions can not be guaranteed with bond-controlled techniques using DSBs or RSBs. While using unbonded DSBs or RSBs would mitigate the concentration of damage at plastic hinge regions, bonded regions limit the probability of residual deformation mitigation. Moreover, unbonded columns hardly can achieve the theoretical strength with a negative or zero post-yield stiffness performance in the inelastic stage. On the other hand, due to the promising performance of the innovative rebars (SFCBs), it is highly desirable to further study the effect of bond conditions on the recoverability of columns reinforced with SFCBs.

5. Analytical study on RC columns reinforced with bond-controlled SFCBs

Fig. 1 shows that the inner core ribbed rebar is first looped around with bundles of continuous fibers roving to fill up the space between its ribs, such that bond strength is guaranteed with the outer longitudinal fibers. This figure also shows that there is a spiral distribution of resin ribs along the SFCB. The main function of those ribs is to ensure the bond performance between a SFCB and the surrounding material, such as concrete, allowing SFCBs to achieve the designed level of strength. For instance, ribs with adequately designed spacing and height would lead SFCBs to approach their ultimate strength; however, in case ribs spacing is wide and the height is undersized, rebars may fail to continue to carry additional load until approaching the designed strength level. Besides, slippages of SFCBs at both yielding and maximum achieved strength depend on the applied bond technique. Consequently, five cases of RC columns reinforced with SFCBs are studied (Table 4), where four cases examine the effect of the achieved strength of SBFCBs on the performance of RC bridge columns. The last case (case no. 5) considers the effect of additional slippage of SFCBs when the ultimate strength is greatly affected by bond conditions, i.e. merely 70% of the ultimate strength could be achieved.

Case	Bond properties			
	S_y (mm)	σ_y (MPa)	S_u (mm)	σ_u (MPa)
1	S_y	f_y	S_u	f_u
2	S_y	f_y	S_u	$0.9f_u$
3	S_y	f_y	S_u	$0.8f_u$
4	S_y	f_y	S_u	$0.7f_u$
5	$2\,S_y$	f_y	$2S_u$	$0.7f_u$

Table 4. Studied bond properties of column reinforced with SFCBs

The studied column has square cross-section of 300x300mm and 1300-mm-tall. It is typified CS10-B30 and reinforced with twelve SBFCB, Table 5. The concrete compressive strength is ~38 MPa. The columns are analyzed under the effect of constant axial load (12% of column axial strength) and reversed cyclic lateral load. Under normal bond condition, a column with these given details was experimentally tested by Fahmy et al. (2010) with another column typified CS10-C40, where its longitudinal reinforcement was twelve SCFCB, Table 5.

Type	Diameter (mm)	Elastic modulus (E_1) (GPa)	Yield strength (MPa)	Post-yield stiffness (E_2),(GPa)	Ultimate strength (MPa)	E_2/E_1 (%)
Φ10+30B						
	13.2	142	312.4	16.6	691.3	11.7
(SBFCB)						
Φ10+40C						
	12.9	155.5	342.2	30.2	641.8	19.7
(SCFCB)						
Φ10	10	200	420	-	480	-

Table 5. Properties of SFCB and ordinary rebars

Figure 19 shows the effect of bond properties on both column post-yield stiffness (its slop and end point) and residual deformations. For instance, with the assumption of extremely sufficient bond strength between SFCBs and the adjoining concrete, Figure 19 (a) shows the highest achieved inelastic stiffness which ends at lateral drift of 40-mm (due to rupture of the outer basalt fibers) with the minimum residual deformations. But, when the composite behavior between SBFCBs and surrounding concrete can not assure the development of the ultimate strength of SFCBs, the inelastic hysteretic response after column yielding would be divided to two parts: the first shows that the column can still carry the load after the SBFCBs yield and hardening behavior has been exhibited; and the second part demonstrates a zero post-yield stiffness, which starts by the end of the hardening zone. As is clear from Figs. 19 (a-e), the slop of the post-yield stiffness and its end point is dependent on the bond conditions. In addition, the controlled increase in slippage of SFCBs would assist in further mitigation of residual deformations, Figs. 19 (d & e).

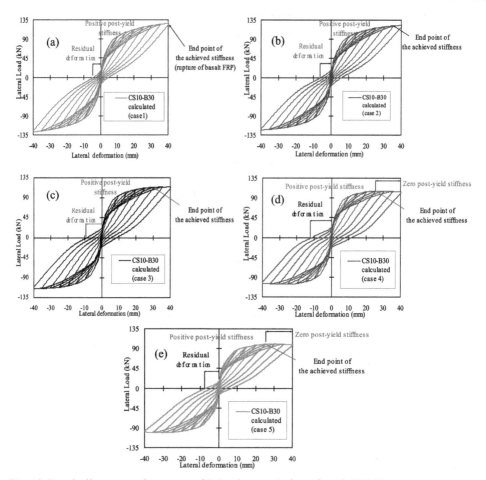

Fig. 19. Bond effect on performance of RC column reinforced with SBFCBs

6. Summary and conclusions

The demand for efficient and effective damage-controllable systems has received strong attention in the last decade, where the main goal is the limitation of the damage at plastic hinge zones along with substantial mitigation of the static residual deformations. Since bond between longitudinal reinforcement and concrete is a key factor controlling structural performance of reinforced concrete (RC) structures, here a study was conducted to examine the effect of alternating the bond conditions between longitudinal reinforcement and concrete on the performance of RC columns. Effects of different conditions of concrete-to-steel bond on the elastic and post-yield stiffnesses, residual deformations, and damage level of RC bridge columns reinforced with rebars having different strain-hardening levels were determined. For ordinary steel bars RSBs and DSBs, the study showed that damage of plastic hinge zone is mitigated provided that reinforcement is completely unbonded from the surrounding concrete. Also, the study revealed that residual deformation of columns reinforced with RSBs can be mitigated only if the unbonded zone represents most of the column height. Despite these encouraging findings, unbonded columns could barely achieve the theoretical strength with zero or negative post-yield stiffness in the inelastic stage. That is, quick recovery of of RC columns reinforced with unbonded DSBs or RSBs can not be guaranteed, which limits the practical application of this technique in RC bridge columns lying in high seismicity zones. Since RC column reinforced with SBFCBs distinctly outperforms its RC counterpart, bond-based damage-controllable system using SBFCBs could be applied for structures reinforced with the innovative rebars, where slope and end point of the achieved post-yield stiffness can be controlled. In addition, both residual deformations and damage level at plastic hinge zone could be limited.

7. References

ACI Committee 318. Building code requirements for structural concrete (ACI 318-08) and commentary (318R-08). *American Concrete Institute (ACI) 2008*, Farmington Hills, Mich., 430 pp.

Cosenza, C.; Manfredi, G. & Ramasco, R. (1993). The use of damage functionals in earthquake engineering: a comparison between different methods. *Earthquake Eng. Struct. Dyn.*, Vol. 22, pp. 855-868.

Fahmy, M.F.M.; Wu, Z.S., Wu, G. & Sun, Z.Y. (2010). Post-yield stiffnesses and residual deformations of RC bridge columns reinforced with ordinary rebars and steel fiber composite bars. *Journal of Engineering Structures*, Vol.32, pp. 2969-2983.

Ghobara, A.; Abou-Elfath, H. & Biddah, A. (1999). Response-based damage assessment of structures. *Earthquake Eng. Struct. Dyn.*, Vol. 28, pp. 79-104.

JSCE Earthquake Engineering Committee (2000). Earthquake resistant design codes in Japan. *Japan Society of Civil Engineers (JSCE)*, Tokyo, Japan, 150 pp.

Kawashima, K.; Hosoiri, K.; Shoji, G., & Sakai, J. (2001). Effects of unbonding of main reinforcements at plastic hinge region on enhanced ductility of reinforced concrete bridge columns. *Structural and Earthquake Engineering, Proc. JSCE, 689/I-57*, pp. 45-64.

Mazzoni, S; McKenne, F; Scott, MH; Fenves, GL, et al. Open System for Earthquake Engineering Simulation User Manual version 2.1.0. Pacific Earthquake Engineering Center, University of California, Berkeley, CA, http://opensees.berkeley.edu/OpenSees/manuals/usermanual/

Moustafa, A. (2011). Damage-based design earthquake loads for single-degree-of-freedom inelastic structures. *Journal of Structural Engineering (ASCE)*, Vol. 137, No.3, pp. 456-467.

Pandey, GR & Mustsuyoshi, H. (2005). Seismic performance of reinforced concrete piers with bond-controlled reinforcements. *ACI Structural Journal*, Vol.102, No.2, pp. 295-304.

Park, R. & Paulay, T. (1975). Reinforced concrete structures. *John Wiley and Sons*, 769pp.

Pettinga, D. ; Pampanin, S. ; Christopoulos, C. & Priestley, N. (2006). Accounting for residual deformations and simple approaches to their mitigation. *First European Conference on Earthquake Engineering and Seismology*, Geneva, Switzerland.

Priestley, M.J.N.; Calvi, G.M. & Kowalsky. (2007). Displacement-based seismic design of structures. *IUSS Press, Pavia*, 721pp.

Saiidi, M.S. ; O'Brein, M. & Sadrossadat-Zadeh, M. (2009). Cyclic response of concrete bridge columns using superelastic nitinol and bendable concrete. *ACI Structural Journal*; Vol.106, No.1, pp. 69-77.

Spacone, E. ; Filippou, F. & Taucer, F. (1996). Fiber beam-column model for nonlinear analysis of R/C frames-part II: applications. *Journal of Earthquake Engineering and Structural Dynamics*, Vol. 25, pp. 727-742.

Takiguchi, K. ; Okada, K. & Sakai, M. (1976). Ductility capacity of bonded and unbonded reinforced concrete members. *Proc. Architectural Institute of Japan.* 249, pp. 1-11.

Wu, G. ; Wu, ZS. ; Luo, YB. ; Sun, ZY. & Hu, XQ (2010). Mechanical properties of steel fiber composite bar (SFCB) under uniaxial and cyclic tensile loads. *Journal of Materials in Civil Engineering*, Vol. 22, No. 10, pp. 1056-1066.

Zhao, J. and Sritharan, S. (2007). Modeling of strain penetration effects in fiber-based analysis of reinforced concrete structures. *ACI Structural Journal*, Vol. 104, No.2, pp. 133-141.

9

Response of Underground Pipes to Blast Loads

A.J. Olarewaju, R.N.S.V. Kameswara and M.A. Mannan
Universiti Malaysia Sabah
Malaysia

1. Introduction

Underground structures are divided into two major categories, fully buried structures and partially buried structures regardless of the shape of the structure. Underground cylindrical structures like pipes, shafts, tunnels, tanks, etc. are used for services such as water supply, sewage, drainage, etc. Most structures have now become targets of terrorist attack in recent years. Examples are 1995 Paris subway in France, 2004 Moscow subway is Russia (Dix, 2004; Huabei, 2009), 1995 Alfred Murrah Federal Building in Oklahoma City. The main sources of blast are: terrorist attacks, war, accidental explosion from military formations, etc. The constituents of blast comprises of: 1) rock media, 2) soils, 3) structure, 4) thin-layer elements surrounding the structure; blast loads, and 5) procedure for the analysis of interaction and responses of these constituents. In order to synchronize the interaction and responses of these variables, relevant data is required which could be obtained from field tests, laboratory tests, theoretical studies, work done in related fields and extension of work done in related fields (Ngo et al., 2007; Greg, 2008; Bibiana & Ricardo, 2008; Olarewaju at al. 2010a).

There are lots of methods available to determine the responses of underground structures to blast loads. These are: i) the analytical methods, and ii) the numerical methods using numerical tools (Ngo et al., 2007; Peter & Andrew, 2009). The problem of analytical method is that the solution allows only a small elastic response or limited plastic response and does not allow for large deflection and may lead to unstable responses. To overcome these problems, the finite element analysis paves the way towards a more rational blast resistance design. Though the drawback is the time and expertise required in pre- and post-processing for a given structural system. In structural design, the methods of structural analysis and design are broadly divided into three categories, namely, theoretical methods which can be used to carry out analysis and the use of design codes, by testing the full size structure or a scaled model using experimental methods, and by making use of model studies (Ganesan, 2000). There are different types of static and dynamic loads acting on underground pipes. In the case of static loads, there surcharge load on the ground surface due various engineering activities. In the case of dynamic loads, these are cyclic load, earthquake, blast, etc. Blast being one of the dynamic load acting on underground pipes either from surface blast, underground blast, open trench blast or internal explosion is a short discontinuous event.

2. Background study

Under blast loading, though typically adopted constitutive relations of soils are elastic, elasto-plastic, or visco-plastic, the initial response is the most important (Huabei, 2009). It involves some plastic deformation that takes place within the vicinity of the explosion and as a result of this one could model the soil as an elasto-plastic material. Beyond this region, the soil can be taken as an elastic material at certain distance from the explosion. Visco-elastic soils exhibit elastic behavior upon loading followed by a slow and continuous increase of strain at a decreasing rate (Duhee et al., 2009). In this study, the soil and pipes are considered as linear elastic, homogeneous, isotropic materials (Boh et al., 2007; Greg, 2008). For such materials, Kameswara (1998) has shown that only two elastic constants are needed to study the mechanics/behavior. These can be the usual elastic constants (the Young's modulus, E and Poisson's ratio,v) or the Lame's constants (λ and μ).

When explosion occurs, surface waves and body waves are generated. Consequent upon these are the isotropic component and deviatory component of the stress pulse (Robert, 2002). Transient stress pulse due to isotropic components causes compression and dilation of soil or rock with particle motion which is known as compression or P-waves. The deviatory component causes shearing of stress with particle velocity perpendicular in the direction to the wave propagation and these are known as shear or S-waves. On the surface of the ground, the particles adopt ellipse motion known as Rayleigh waves or R-waves (Kameswara Rao, 1998; Robert 2002). Energy impulse from explosion decreases for two reasons: (i) due to geometric effect, and (ii) due to energy dissipation as a result of work done in plastically deforming the soil matrix (Dimitiri & Jerosen, 1999; Huabei, 2009; Omang et al., 2009).

The categories of blast in this study that are applicable to underground pipes are; (i) underground blast, (ii) blast in open trench, (iii) internal explosion inside the pipes as well as (iv) surface blast (Olarewaju et al. 2010b). Blasts can create sufficient tremors to damage substructures over a wide area (Eric Talmadge and Shino Yuasa, 2011). With regards to the severity of destruction of explosion as a result of blast, it has been reported by James (2008) that typical residence structure will collapse by an overpressure of 35 kPa while a blast wave of 83 kPa will convert most large office buildings into rubbles. Accordingly, blast could be thought of as an artificial earthquake. Consequently, there is need to study the relationships and consequences of blasts in underground structures specifically in pipes. This is with a view to designing protective underground structures specifically pipes to resist the effects of blast and to suggest possible mitigation measures.

A lot of works have been done on dynamic soil-structure interaction majorly for linear, homogeneous, and semi-infinite half space soil media. This is contained in Olarewaju et al. (2010a). In this work, observations were limited to displacements at the crown and spring-line of pipe buried in a soil layer. Effect of slip between the soil and pipe was not considered. Huabei (2009) recently obtained the responses of subway structures under blast loading using the Abaqus finite element numerical software. This study is limited to the determination of the responses of empty underground pipes under blast loads. The material properties are limited to linear, elastic, homogeneous and isotropic materials. It is assumed that blast takes place far away from the vicinity of the underground pipes.

3. Blast load characteristics and determination

Explosive has to detonate in order to produce explosive effect. The term detonation as explained in the Unified Facilities Criteria (2008) refers to a very rapid and stable chemical reaction that proceeds through the explosive material at a speed termed the detonation velocity. This velocity ranges from 6705.6 m/s to 8534.4 m/s for high explosives. The detonation waves rapidly convert the explosive into a very hot, dense, high-pressure gas.

The volume of the gas of this explosive material generates strong blast waves in air. The pressures behind the detonation front range from 18619 MPa to 33785 MPa. Only about one-third of the total energy generated in most high explosives is released in the detonation process. The remaining two-thirds of the energy is released in air more slowly during explosions as the detonation products mix with air and burn.

According to the same source, the blast effects of an explosion are in the form of shock waves composed of high-intensity shocks which expand outward from the surface of the explosive into the surrounding air. As the shock wave expand, they decay in strength, lengthen in duration, and decrease in velocity (Longinow & Mniszewski, 1996; Remennikov, 2003; Unified Facilities Criteria, 2008). According to the Unified Facilities Criteria (2008), blast loads on structures can be categorized into two main headings; i) unconfined explosions (i. e. free air burst, air burst and surface), ii) confined explosions (i. e. fully vented, partially confined and fully confined).

According to the same source, the violent release of energy from a detonation converts the explosive material into a very high pressure gas at very high temperatures. This is followed by pressure front associated with the high pressure gas which propagates radially into the surrounding atmosphere as a strong shock wave, driven and supported by the hot gases. The shock front, term the blast wave is characterized by an almost instantaneous rise from atmospheric pressure to a peak incident pressure Pso. Over pressure, Pso is the rise in blast pressure above the atmospheric pressure. This pressure increases or the shock front travels radially from the point of explosion with a diminishing shock velocity U which is always in excess of the sonic velocity of the medium. The shock front arrives at a given location at time t_A (ms). After the rise to the peak value of over pressure Pso, the incident pressure decays to the atmospheric value in time t_o (ms - millisecond) which is the positive duration (Olarewaju et al. 2011n).

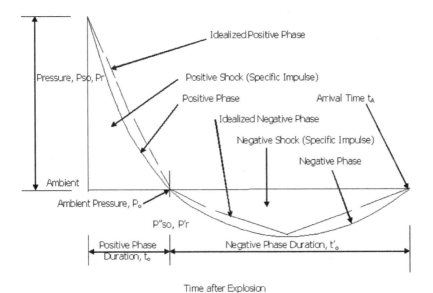

Fig. 1. Pressure Time Variation (Unified Facilities Criteria, 2008; Olarewaju et al.2011 and 2011n)

The negative phase with duration t0⁻ (ms) is usually longer than the positive phase. It is characterized by a negative pressure (usually below atmospheric pressure) having a maximum value of negative overpressure Pso⁻ as well as reversal of the particle flow. The negative phase is usually less important in design than the positive phase because it is very small and is usually ignored. The incident pulse density (i. e., specific impulse) associated with the blast wave is the integrated area under the pressure-time curve and is denoted by i_s for the positive phase and by $i_s⁻$ for the negative phase as illustrated in Fig. 1. An additional parameter of the blast wave, the wave length, is sometimes required in the analysis of structures. The positive wave length $L_W⁺$ is the length at a given distance from the detonation which, at a particular instance of time, is experiencing positive pressure (Longinow & Mniszewski, 1996; Remennikov, 2003; Unified Facilities Criteria, 2008). Unified Facilities Criteria (2008) allows for an increase of 20%.

In case of underground blast, most of the energy is spent in fracturing, heating, melting, and vaporizing the surrounding soils and rocks (Johnson & Sammis, 2001) with only a very small amount being converted to seismic energy. The fraction of the small amount of total energy that goes into seismic energy is a measure of the seismic efficiency of underground explosions. There are three methods available for predicting blast loads on structures. These are: empirical, semi-empirical and numerical methods. Details could be found in Peter and Andrew (2009), Olarewaju (2010), Olarewaju et al. (2010i), (2010j) and (2011p).

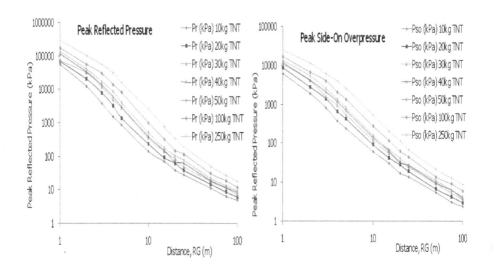

Fig. 2. Peak Reflected Pressure and Peak Side-On Overpressure for Surface Blast (Olarewaju et al. 2010c, 2010e, 2010i)

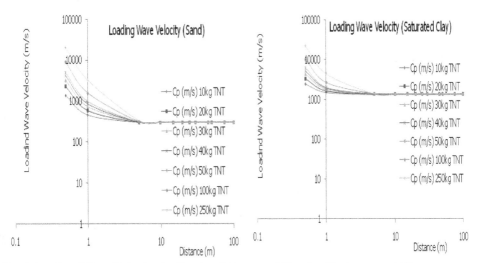

Fig. 3. Loading Wave Velocity for Sand and Saturated Clay for Underground Blast (Olarewaju, et al. 2010c, 2010e, 1020f, 2010i)

Mitigation techniques are meant to reduce the impact of blast and seismic related issues on underground structures. These techniques include: soil stabilization using mechanical and/or additive, grout, ground improvement using i) prefabricated vertical drains, placing soil surcharge and maintaining it for the required time, vacuum consolidation, stone column; ii) chemical modification (with deep soil mixing, jet grouting, etc); iii) densification (using vibro compaction dynamic compaction, compaction grouting, etc), reinforcement (using stone columns, geo-synthetic reinforcement) (Olarewaju, 2004a; 2008b; Raju, 2010; Kameswara, 1998; Olarewaju et al. 2011). Tire-chip backfill has also been used by Towhata & Sim (2010) to reduce the bending stress and moment caused by displacement of underground pipes. If the thickness of the tire-chip backfilling is increased, it can resist larger displacement caused by blast and thereby reduces the bending stress and the moment caused by large displacements. Similarly, trenchless technique can also be used to rehabilitate damaged underground pipes due to blast, aging, etc. (Randall, 1999) especially in congested and built-up areas.

4. Methodology

The existing model studied by Ronanki (1997) was validated using the Abaqus numerical package and the results are compared well. From the results, the crown displacement at H/D=1 is 1.31 times that of crown displacement at H/D=2. The maximum horizontal sprig-line response in terms of pressure, displacement, maximum principal strain and mises for H/D=1 is 1.24 times that of maximum horizontal spring-line response for H/D=2. This is in line with the submissions of Roanaki (1997) that "Embedment depth has significant effect on both the crown and spring-line response (deflection). With increase of depth of embedment of pipes, the response (deflection) decreases. The maximum crown response for H/D=1 is about 1.3 times that of the maximum crown response (deflection) of H/D=2. In case of spring-line response (deflection), the maximum horizontal spring-line deflection for H/D=1 is about 1.2 times that of maximum horizontal spring-line deflection of H/D=2".These results is also in agreement with those reported by Ramakrishan (1979) though no numerical data are presented.

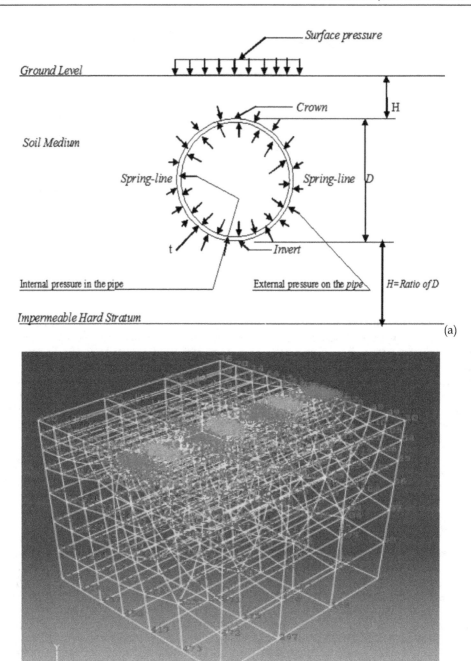

Fig. 4. (a) Cross-section of underground pipe (Olarewaju et al. 2011n); (b) Finite element model of underground pipe using Abaqus

Material	Density, ρ (kg/m³)	Young's Modulus, E (kPa)	Poisson's Ratio, υ
Loose sand	1800	18500	0.3
Dense sand	1840	51500	0.375
Undrained Clay	2060	6000	0.5
Intervening medium	1800	18500	0.3
Steel pipe	7950	200×10^6	0.2
Concrete pipe	2500	20×10^6	0.175

Table 1. Material properties for the study

The ground media considered in this study are loose sand, dense sand and undrained clay. The geotechnical properties shown in Table 1 as revealed by several researchers (Das, 1994; FLAC, 2000; Coduto, 2001; Duncan, 2001; Unified Facilities Criteria, 2008; Kameswara, 1998; etc) were used to study the response of underground pipes due to blast loads. Since the two elastic constants are enough to study the mechanics of an elastic body, the material properties used are the modulus of elasticity, E, Poisson's ratio and density of soil and pipe materials. The largest possible value of Poisson's ratio is 0.5 and is normally attained during plastic flow and this signifies constancy of volume (Chen, 1995). Huabei (2009) pointed out that undrained behavior is relevant for saturated soft soils especially clay that is subjected to rapid blast loading since the movement of pore water is negligible under such circumstance. For 10kg, 20kg, 30kg, 40kg, 50kg, 100kg and 250kg explosives, Unified Facilities Criteria (2008) was used to predict positive phase of blast loads at various stand-off point for surface blast and results are presented in Figs 2. Analytical method was used to predict the blast load for underground blast at various stand-off points and results presented in Figs. 3. According to Huabei (2009), it is not likely for terrorists to use very large amount of explosive in an attack targeting underground pipes. Soil model in the problem definition shown in Figs. 4 (a, b) of 100m by 100m by 100m depth consist of buried pipe 100m long and 1m diameter buried at various embedment ratios were study for the various categories of blast applicable to underground pipes. Parametric studies were carried out for various blasts. Blast load duration was verified and it was observed that, for response to take place in underground pipe, most especially pipes buried in loose sand, duration of blast should be greater than 0.02s (Olarewaju, et al 2011n).

5. Method of analysis

Abaqus package was used to solve the equations of motion of the system:

$$[m] [\ddot{U}] + [c] [\dot{U}] + [k] [U] = [P] \tag{1}$$

with the initial conditions:

$$U (t = 0) = U_o \text{ and } \dot{U} (t = 0) = \dot{U}_o = v_o \tag{2}$$

where m, c, and k are the global mass, damping and stiffness matrices of the pipes system and t is the time. U and P are displacement and load vectors while dot indicate their time derivatives. The time duration for the numerical solution (Abaqus Analysis User's Manual, 2009) was divided into intervals of time $\Delta t = h$, where h is the time increment. Finite difference in Abaqus/Explicit was used to calculate the response (Abaqus Analysis User's

Manual, 2009). Stability limit is the largest time increment that can be taken without the method generating large rapid growing errors (Abaqus Analysis User's Manual, 2009; Abaqus/Explicit: Advanced Topics, 2009). The difficulty is that the accuracy of the sensitivities can depend on the number of elements. This dependency is not seen with either analytical sensitivity analysis or with the overall finite difference method (explicit). Sensitivity analysis is not required in Finite difference of Abaqus/Explicit because. According to Abaqus Analysis User's Manual (2009), the default value of perturbation has been proved to provide the required accuracy in Abaqus /Standard.

Boundary condition of the model was defined with respect to global Cartesian axes in order to account for the infinite soil medium (Geoetchnical Modeling and Analysis with Abaqus, 2009; Ramakrishan, 1979; Ronanki, 1997). Contrary to our usual engineering intuition, introducing damping to the solution reduces the stable time increment. However, a small amount of numerical damping is introduced in the form of bulk viscosity to control high frequency oscillations (Abaqus Analysis User's Manual, 2009; Geoetchnical Modeling and Analysis with Abaqus, 2009). Estimation of blast load parameters could be done by empirical method, semi-empirical methods and numerical methods. The method to be adopted depends on the numerical tool available for the study of response of underground structures to blast loads. In this study, empirical method using Unified Facilities Criteria (2008) was used. The blast load parameters to be determined using this method depend on the available numerical tool. According to Unified facilities Criteria (2008), pressure is the governing factor in design and the study of the response of underground structures. Load due to surface blast was represented by pressure load with short duration (in millisecond, ms) while load due to underground blast was represented by loading wave velocity with short load duration (in millisecond, ms).

6. Results and discussion

6.1 Response of underground pipes to surface blast

The blast load was represented by pressure load (Figures 1 and 2) whose centre coincide with the centre of the explosive. The pressure load reduces to zero at 0.025s. At low pressure load due surface blast, there was no response observed on underground pipe. Due to surface blast, it was observed that crown, invert and spring-line displacement reduces as embedment ratios increases in loose sand, dense sand and undrained clay. This is shown shown in Figs. 5. Crown, invert and spring-line pressures, stresses and strains increase at embedment ratios of 2 and 3 after which it reduces as the embedment ratios increases.

For steel pipe at $H/D = 1$, crown and invert displacement in loose sand is the highest and least in undrained clay. This is in agreement with the findings of Huabei (2009), that increasing the burial depth enhances the confinement of underground pipe, hence reduces the maximum lining stress under internal blast loading (Huabei, 2009). The results indicate that it is necessary to evaluate the blast-resistance of underground structures with small burial depth. Materials yield easily and more at lower depth of burial (Huabei, 2009).

With small burial depth, due to low confinement from ground, displacement, pressure, stress and strain could be significantly large and underground structures like pipes could be severely damaged even with moderate surface blast, underground blast and open trench blast (Olarewaju et al 2010c). According to James (2009), the effect of varying the depth of burial of structures below the ground level is an important phenomenon to study. The depth of soil cover above the increases the over burden stresses on it, which can help in stabilizing it with respect to its response to an externally applied impulsive action. This can help in reduction of the vibrations which occur in response to an explosive blast action.

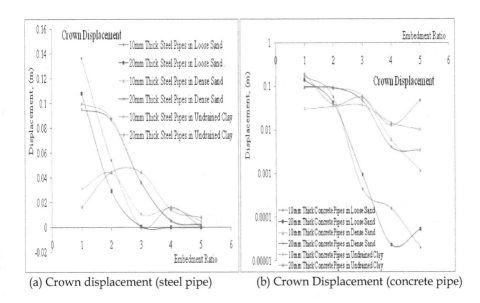

(a) Crown displacement (steel pipe)   (b) Crown Displacement (concrete pipe)

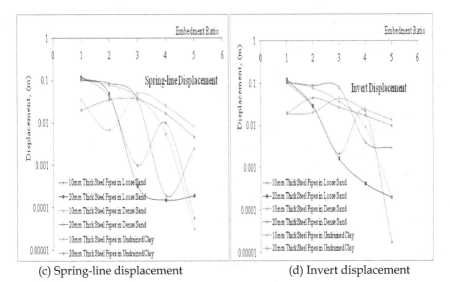

(c) Spring-line displacement   (d) Invert displacement

Fig. 5. Displacement in underground pipes due to surface blast

6.2 Response of underground pipes to underground blast

The blast load was represented by loading wave velocity (Figures 1 and 3) which reduces to zero at 0.025s. For a given loading wave velocity, crown, invert and spring-line displacements in pipes is almost constant at all the embedment ratios considered irrespective of the material properties. This is higher compared to that obtained in open trench blast. This is because, as the peak particle velocity due to underground blast travels within the soil medium, it transmits the load bodily to the buried pipes along the direction of travel. As a result of this, displacement is bound to be higher compared to open trench blast where the wave energy only impeaches on the side of the trench.

Reduction in pressure, stress and strain is noticeable at embedded ratios of 3 to 5. This is in agreement with the submission of Ronanki (1997) on the effect of seismic/loading wave velocity that, the spring-line horizontal displacement remains almost constant with increasing mode shape number. The vertical crown displacement increases with mode shape number up to a value 15, beyond that the displacement tends to be constant (Ronanki, 1997). Finally, crown, invert and spring-line pressures, stresses and strains in pipes showed wide variation as the embedment ratio increases in all the soil media considered. Though there is reduction in all these parameters as the embedment ratio increases (Olarewaju et al. 2010f).

6.3 Response of underground pipes to open trench blast

The blast load was represented by pressure load (Figures 1 and 2) which reduces to zero at 0.025s. Displacement (Figs. 6) in pipes due to open trench blast is lower compared to that obtained in underground blast. In addition, virtually all the parameters observed i. e. displacement, pressure, stress and strain at the crown, invert and spring-line of pipes reduces at embedment ratios of 3 beyond which no significant changes occurred. Finally, crown, invert and spring-line displacements, pressures, stresses and strains reduce as the embedment ratio increases with a sharp increase at embedment ratio of 2 in all the ground media considered (Olarewaju et al. 2010e). Increasing the burial depth of underground pipe enhances the confinement on the underground pipe, hence reduces the maximum displacement, pressure, stress and strain under blast loading (Huabei, 2009). Details could be found in Olarewaju et al (2010b)

6.4 Response of underground pipes to internal explosion

The blast load was represented by pressure load (Figures 1 and 2) whose centre coincide with the centre of the explosive. The pressure load reduces to zero at 0.025s. The result shows that as the diameter of pipes increases, blast load parameters generated inside the pipe increases. As the thickness of pipes reduces, time history as a result of internal explosion increases in the same proportion. In addition to this, depth of burial of pipes showed no significant changes in the time history of external work and energies generated due to internal explosion (Olarewaju et al 2010d and 2010l). Furthermore, stress components on the ground surface reduced as the depth of embedment of pipes increases. Equivalent earthquake parameters on the surface of the ground due to 50kg TNT explosion in pipe are higher than that recorded in San Fernando earthquake of 1971 (Robert, 2002). Finally, pressure changes from negative to positive within the soil medium due to dilations and compressions caused by the transient stress pulse of compression wave while velocity, displacement and stresses reduce as it approaches the ground surface. This reduction is more in loose sand than undrained clay due to arching effect (Craig, 1994). Details could be found in Olarewaju et al. (2010d).

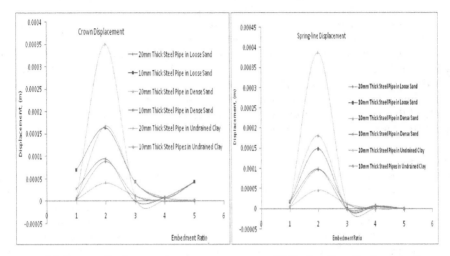

(a) Crown displacement   (b) Spring-line displacement

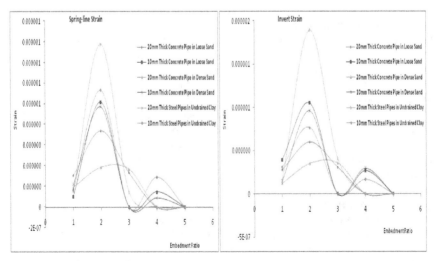

(c) Spring-line strain   (d) Invert strain

Fig. 6. Displacement and Strain in pipes in open trench blast

7. Parametric studies

7.1 Effects of coefficient of friction

Due to surface blast, displacement at the crown reduces at coefficient of friction of 0.2 to 0.4 and above in dense sand. The reverse is the case in loose sand where displacement increases as the coefficient of friction increases. Invert displacement reduces as the coefficient of friction increases. Spring-line displacement increases as the coefficient of friction increases. Due to the dynamic nature of surface blast loads, there is wide variation in the results; there is reduction in the values of crown, invert and spring-line pressures, stresses and strains for coefficient of friction of 0.2 to 0.4. This is also noticeable for the increased values of peak reflected pressure. Liang-Chaun (1978) pointed out that in cases when test data are not available, the following friction coefficient can be used: Silt = 0.3; Sand = 0.4; Gavel = 0.5m and added that the above coefficients are the lower bond values equivalent to the sliding friction. The static and dynamic coefficient of friction can be as much as 70% higher.

7.2 Effects of young's modulus of soil

Effects of liquefaction as observed in the varying Young's modulus for soil for surface blast and underground blast is similar to the varying Young's modulus for intervening medium. Varying the Young's modulus, E of soil, displacement became higher at E of 1×10^6 Pa. Between Young's modulus, E of 10000 Pa and 3000000 Pa, pressure, stress and strain get to the peak value with maximum value at E of 1000000 Pa. Crown has the maximum values of stress and strain while invert has the maximum pressure. With the value of Young's modulus, E soil ranging from 0 Pa to 10000 Pa, displacement, pressure, stress and strain (Figs 7) reduce with no substantial increase. This is in agreement with the submission of Susana & Rafael (2006). From the result of the work by Huabei (2009), it showed that as Young's modulus of soil is increasing, mises stress is reducing. For the constant value of stress with increasing value of Young's modulus, E of soil, it shows that the soil has yielded.

7.3 Effects of young's modulus of pipes

Displacement is high at the crown but low at the invert and spring-line of pipes having low value of Young's modulus. At higher Young's modulus, the displacement at the crown, invert and spring-line became the same. Pressure and stress is low at low Young's modulus but increases as the Young's modulus increases. Large strain is observed between the values of 100Pa and 10000Pa beyond which the value of strain reduces. Low stiffness pipes are pvc pipes, clay pipes, etc while high stiffness pipes are steel pipes, reinforced concrete pipes, etc. It is evident that as the Young's modulus E of pipes increases, strain reduced due to increased stiffness but the pressure and stress increases from E of 1×10^7 Pa. This shows that pipes of lower value of E have lower displacement, pressure, stress and strain induced in them due to surface blast compared to pipes with higher stiffness like steel and reinforced concrete pipes.

The result presented by Frans (2001) clearly shows that the low stiffness pipes suffer less from subsidence than the one with the higher stiffness. At the same time a higher deflection is observed when using low stiffness pipes. This proves that rigid pipes transfer load, and flexible pipes deform and the load is transferred by the soil. When the bed is firm, hardly any subsidence takes place hence the stiffness of the pipe has no effect either. However, when the bed is loose or soft, subsidence becomes a real issue and also the effect of pipe stiffness is significant.

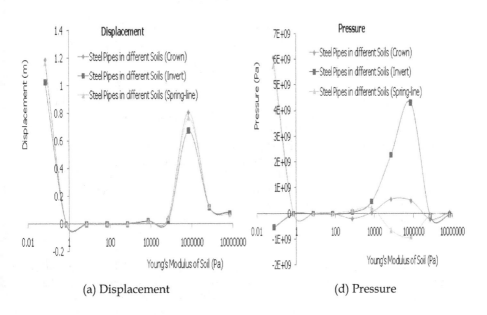

(a) Displacement   (d) Pressure

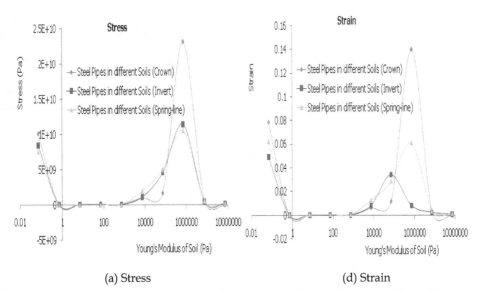

(a) Stress   (d) Strain

Fig. 7. Displacement, Pressure, Stress and Strain in buried pipes for varying Young's modulus of soil for surface blast

7.4 Effects of young's modulus of pipes

Displacement is high at the crown but low at the invert and spring-line of pipes having low value of Young's modulus. At higher Young's modulus, the displacement at the crown, invert and spring-line became the same. Pressure and stress is low at low Young's modulus but increases as the Young's modulus increases. Large strain is observed between the values of 100Pa and 10000Pa beyond which the value of strain reduces. Low stiffness pipes are pvc pipes, clay pipes, etc while high stiffness pipes are steel pipes, reinforced concrete pipes, etc. It is evident that as the Young's modulus E of pipes increases, strain reduced due to increased stiffness but the pressure and stress increases from E of 1×10^7 Pa. This shows that pipes of lower value of E have lower displacement, pressure, stress and strain induced in them due to surface blast compared to pipes with higher stiffness like steel and reinforced concrete pipes.

The result presented by Frans (2001) clearly shows that the low stiffness pipes suffer less from subsidence than the one with the higher stiffness. At the same time a higher deflection is observed when using low stiffness pipes. This proves that rigid pipes transfer load, and flexible pipes deform and the load is transferred by the soil. When the bed is firm, hardly any subsidence takes place hence the stiffness of the pipe has no effect either. However, when the bed is loose or soft, subsidence becomes a real issue and also the effect of pipe stiffness is significant.

7.5 Effects of pipe thickness

The result indicates that steel and concrete pipes show similar characteristics and behavior in thickness. In other words, as the thickness of pipes increases, displacement, pressure, stress and strain reduces. At low pipe thickness, displacement, stress and strain in steel and concrete pipes buried in undrained clay, is low at the invert but remain constant at the crown, invert and spring-line as the thickness increases. According to James (2009), the size and thickness of the structure under consideration is a major factor which can potentially influence the stresses generated on it. The reason could be attributed to the fact that smaller size structure has lower mass, making it easier to displace under blast loadings.

Higher displacements in the structure can result in larger strain deformations, causing the corresponding stresses to be lower due to energy dissipation in deforming the structure. According to Zhengwen (1997), rigid structures experience higher pressure and less displacement during the first half-wave of response, when compared with more flexible counterparts. In that case, underground pipes with smaller thickness are considered as flexible while those with increased thickness are considered as rigid structures.

8. Conclusions

Blast is a short discontinuous event whose duration is very small compared to earthquake. Considering the various constituent of blast, ground pipes and intervening media can be modeled. It must be remembered that soil exists as semi-infinite medium. Numerical tool to be used must incorporate the notion of infinite in the formulation. To account for the infinity of soil medium, in this study, in the absence of infinite element, Global Cartesian axis in Abaqus software was used. In other words, it shows that soil is a continuous media. To account for material damping, small numerical damping in the form of bulk viscosity was introduced. Blast and/or blast parameters can be represented or modeled using software (i. e. BLASTXW, SPLIT-X, BLAPAN, SPIDS, etc) or by using Eulerian numerical techniques developed using finite volume and finite difference solver (i. e. SHAMRC, ANSYS, AUTODYN 2D AND 3D, etc) (Olarewaju et al., 2010i). To represent blast load parameters, it can to be determined by empirical method using available code like Technical Manual 1990, Unified Facilities Criteria 2008, etc (Unified Facilities Criteria 2008 supersede other available technical manual). In this study, blast load parameters were estimated using empirical method, (i. e. Unified facilities Criteria (2008)) and represented in the model. Other blast load parameters applicable to the design and study of response underground pipes to blast loads that can be estimated by empirical method are: peak reflected pressure, side-on overpressure, specific impulse, horizontal and vertical acceleration, horizontal and vertical displacement, shock front velocity, horizontal and vertical velocity, duration, arrival time, etc (Olarewaju et al. 2010a; 2010i). To capture the short duration of blast load, time integration technique in Abaqus/Explicit was used in this study.

Conclusively, this study has shown the various responses of underground pipes due to various blasts scenarios. Results of parametric studies were also presented and discussed. Finally, possible mitigation measures were also suggested. Consequently, the parameters thus obtained will help in designing underground pipes to resist effects of various blast loads.

9. Acknowledgment

The financial supports provided by Ministry of Science Technology and Innovation, MOSTI, Malaysia under Universiti Malaysia Sabah (UMS) e-Science Grant no. 03-01-10-SF0042 is gratefully appreciated.

10. References

Abaqus Inc. (2009). Abaqus Analysis User's Manuals - Documentation, Dassault Systemes Simulia, Providence, Rhode Island, USA

Abaqus Inc. (2009). Geotechnical Modeling and Analysis with Abaqus, Dassault Systemes Simulia, Providence, Rhode Island, USA

Bibiana, M. Luccioni, and D. Ricardo Ambrosini, (2008). Evaluating the effect of underground explosions on structures, Asociacion Argentina de Mecanica Computacional, Mecanica Computacional XXVII, Noviembre, pp 1999-2019

Boh, J. W., Louca, L. A. and Choo, Y. S. (2007). Finite Element Analysis of Blast Resistance Structures in the Oil and Gas Industry, Singapore and UK, ABAQUS User's Conference, pp 1-15

Chen, W. F. (1995). The Civil Engineering Handbook, CRC Press, London. 1386 Coduto, D. P. (2001). Foundation Design: Principles and Practices (2nd edition), Prentice Hall, Inc., New Jersey, 883

Craig, R. F. (1994). Soil Mechanics, (5th edition), Chapman and Hall, Great Britain Das, B. M. (1994). Principles of Geotechnical Engineering (3rd edition), PWS Publishing, Co., Boston, Massachusetts, 672

Dix, A. (2004). Terrorism – the new challenge for old tools. Tunnels and Tunneling International 36 (10): pp 41-43

Duncan, J. M. (2001). CEE 5564: Seepage and Earth Structure, Course Notes, Spring 2001, Virginia Tech, Blacksburg, Virginia

Dimitri Komatitsch and Jerosen Tromp (1999). Introduction to the spectral element method for three dimensional seismic wave propagation, Geophys. J. Int., 139, pp 806-822

Duhee P., Myung S., Dong-Yeop K. and Chang-Gyun J., (2009). Simulation of tunnel response under spatially varying ground motion, International Journal Soil Dynamics and Earthquake Engineering, Science Direct Ltd Eric Talmadge and Shino Yuasa, 2011. Stricken Japan nuclear plant rocked by 2nd blast.

Fukushima Dai-Ichi Nuclear Plant Plagued By Cooling Issues The Associated Press, 14th March. http://www.news4jax.com/nationalnews/27184574/detail.html

FLAC User's Manual (2000). Version 4.0. Itasca Consulting Group, Inc., Minneapolis, Minnesota

Frans Alferink. (2001). Soil-Pipe Interaction: A next step in understanding and suggestions for improvements for design methods, Waving M & T, The Netherlands, Plastic Pipes XI, Munich, 3rd-6th September

Ganesan, T. P. (2000). Model of Structures (1st edition), University Press Ltd., India, ISBN: 817371 123-2

Greg B. C. (2008). Modeling Blast Loading on Reinforced Concrete Structures with Zapotec,Sandia National Laboratories, Albuquerque, ABAQUS User's Conference

Huabei Liu, (20090. Dynamic Analysis of Subways Structures under Blast Loading,University Transportation Research Center, New York, USA

James A. Marusek, (2008). Personal Shelters, Abaqus User's Conference, US Department ofthe Navy

Johnson, L. R. and Sammis, C. G. (2001). Effects of rock damaged on seismic wavesgenerated by explosions. International Journal of Pure Applied Geophysics, 158, pp 1869-1908

Kameswara Rao, N. S. V. (1998). Vibration Analysis and Foundation Dynamics (1st edition),Wheeler Publishing Co. Ltd., New Delhi, India, ISBN: 81-7544-001-5

Liang-Chaun Peng. (1978). Soil-pipe interaction - Stress analysis methods for undergroundpipelines, AAA Technology and Specialties Co., Inc., Houston, Pipeline Industry, May, 67-76

Longinow, A. and Mniszewski, R. K. (1996). Protecting buildings against vehicle bombattacks, Practice Periodical on Structural Design and Construction, ASCE, New York, pp 51-54

Mosley, W. H. and Bungey, J. H. (1990). Reinforced Concrete Design (4th edition), ELBS, withMacmillan, Hampshire

Neil, J. and Ravindra, K. D. (1996). Civil Engineering Materials, (5th edition), MacmillanPress Ltd., London

Ngo, T. J., Mendis, J., Gupta, A. and Ramsay, J. (2007). Blast Loading and Blast Effects onStructures – An Overview, University of Melbourne, Australia, International Journal of Structural Engineering, EJSE, International Special Issue: Loading on Structures, pp 76-91

Olarewaju, A. J., Kameswara Rao, N.S.V and Mannan, M.A., (2010a). Response ofUnderground Pipes due to Blast Loads by Simulation – An Overview, Journal of Geotechnical Engineering, EJGE, (15/G), June, pp 831-852, ISSN 1089-3032

Olarewaju, A. J., Kameswara Rao, N.S.V and Mannan, M.A., (2010b). Guidelines for theDesign of Buried Pipes to Resist Effects of Internal Explosion, Open Trench and Underground Blasts, Journal of Geotechnical Engineering, EJGE, (15/J), July, pp 959-971, ISSN 1089-3032

Olarewaju, A. J., Kameswara Rao, N.S.V and Mannan, M.A., (2010c). Blast Effects onUnderground Pipes, Journal of Geotechnical Engineering, EJGE, (15/F), May, pp 645-658, ISSN 1089-3032

Olarewaju, A. J., Kameswara Rao, N.S.V and Mannan, M.A., (2010d), Behaviors of BuriedPipes due to Internal Explosion, Malaysia Construction Research Journal, MCRJ, September, (in press)

Olarewaju, A. J., Kameswara Rao, N.S.V and Mannan, M.A., (2010e), Design Hints forBuried Pipes to Resist Effects of Blast, Proceedings of Indian Geotechnical Conference (IGC), ISBN 13: 978-0230-33211-9, ISBN 10: 0230-33207-2, Indian Institute of Technology, Powai, Mumbai, India, Macmillan Publishers India Ltd., December 16th-18th, Published in 2011, pp 881-884

Olarewaju, A. J., Kameswara Rao, N.S.V and Mannan, M.A., (2010f), Response ofUnderground Pipes Due to Underground Blast, Proceedings of the International Agricultural Engineering Conference on Innovation, Cooperation and Sharing, Chinese Academy of Agricultural Mechanization Sciences and Shanghai Society for Agricultural Machinery, Shanghai, China, September 17th-20th, pp (I) 321-329

Olarewaju, A. J., Kameswara Rao, N.S.V and Mannan, M.A., (2010g), Behavior of BuriedPipes Due to Surface Blast Using Finite Element Method, Proceedings of 1st Graduate Student Research International Conference, Brunei, on Contributions Towards Environment, Bio-diversity and Sustainable Development, Universiti Brunei Darussalam, December 13th-15th, pp 17 (1-6)

Olarewaju, A. J., Kameswara Rao, N.S.V and Mannan, M.A., (2011h), Response ofUnderground Pipes Due to Surface Blast Using Finite Element Method, Proceedings of International Soil Tillage Research Organization (ISTRO) – Nigeria Symposium on Tillage for Agricultural Productivity and Environmental Sustainability, University of Ilorin, Ilorin, Nigeria, February 21st-24th, 241-251

Olarewaju, A. J., Kameswara Rao, N.S.V and Mannan, M.A., (2010i). Blast Prediction andCharacteristics for Simulating the Response of Underground Structures, Proceedings of the 3rd International Conference of Southeast Asian on Natural Resources and Environmental Management (3-SANREM), ISBN: 978-983-2641-59-9, August 3rd-5th, Universiti Malaysia Sabah, Malaysia, pp 384 – 391

Olarewaju, A. J. (2010j). Blast Effects on Underground Pipes, SKTM (School of Engineeringand Information Technology) PG Newsletter, Universiti Malaysia, Sabah, Special Ed. (1/1), July, pp 5, ISSN 2180-0537

Olarewaju, A. J., Kameswara Rao, N.S.V and Mannan, M.A., (2011k), Blast Effects onUnderground Pipes Using Finite Element Method, Proceedings of 12th International Conference on Quality in Research, ISSN: 114-1284, Faculty of Engineering, University of Indonesia, Bali, Indonesia, July 4th-7th, (in press)

Olarewaju, A. J., Kameswara Rao, N.S.V and Mannan, M.A., (2010l), Response ofUnderground Pipes due to Blast Load, Proceedings of the 3rd International Earthquake Symposium Bangladesh, ISBN: 978-984-8725-01-6, Bangladesh University of Engineering Technology, Dhaka, 4th-6th March, pp 165-172

Olarewaju, A. J., Kameswara Rao, N.S.V and Mannan, M.A., (2011m). Dimensionless Response of Underground Pipes Due to Blast Loads Using Finite Element Method, Journal of Geotechnical Engineering, EJGE, (16/E), March, ISSN 1089-3032, pp 563-574

Olarewaju, A. J., Kameswara Rao, N.S.V and Mannan, M.A., (2011n). Simulation and Verification of Blast Load Duration for Studying the Response of Underground Horizontal and Vertical Pipes Using Finite Element Method, Journal of Geotechnical Engineering, EJGE, (16/G), April, ISSN 1089-3032 pp 785-796

Olarewaju, A. J. (2008a). Engineering Properties of Recycled Concrete and Used Steel Reinforcement Bars. M. Eng Thesis, Civil Engineering Department, Federal University of Technology, Alure, Ondo State, Nigeria, July

Olarewaju, A. J., (2004b). Soil Stabilization, M.ENG Seminar Paper on Course Title: Advanced Soil Mechanics (CVE 821), Civil Engineering Department, Federal University of Technology, Akure, Ondo State, Nigeria, June

Olarewaju, A. J., Balogun, M. O. and Akinlolu, S. O. (2011). Suitability of Eggshell Stabilized Lateritic Soil as Subgrade Material for Road Construction. Electronic Journal of Geotechnical Engineering (EJGE), ISSN 1089-3032, (16/H), April, pp 899-908

Omang, M. Borve, S. and Trulsen, J. (2009). Numerical Simulations of Blast Wave Propagation in Underground Facilities, Norwegian Defense Research Establishment, Norway

Peter, D. S. and Andrew, T. (2009). Blast Load Assessment by Simplified and Advanced Methods, Defence College of Management and Technology, Defence Academy of the United Kingdom, Cranfield University, UK. http://www.civ.uth.gr/

Prakash, S. and Puri, V. K. (2010). Past and Future of Liquefaction, Proceedings of Indian Geotechnical Conference, ISBN: 978-0230332089, ISBN: 0230-332080, GEOtrendz, Macmillan Publisher India Ltd., Vol. III, 63-72

Raju, V. R. (2010). Ground Improvement – Application and Quality Control, ISBN: 978 0230332089, ISBN: 0230-332080, Proceedings of Indian Geotechnical Conference, GEOtrendz, Macmillan Publisher India Ltd., Vol. III, 121-131

Ramakrishan, K. (1979). Finite element analysis of pipes buried in linearly and non linear elastic media. M.Tech thesis, Civil Engineering Department, Indian Institute of Technology, Kanpur, India

Randall C. Conner, (1999). Pipeline and Rehabilitation, Proceedings of the group of ASCE Technical Sessions at the 1999 American Public Works Association International Public Works Congress and exposition, September, 19-22, Denver Colorado

Remennikov, A. M. (2003). A Review of Methods for Predicting Bomb Blast Effects on Buildings, University of Wollongong, International Journal of Battlefield Technology, 6(3): pp 5-10

Robert, W. D. (2002). Geotechnical Earthquake Engineering Handbook, McGraw Hill, New York, ISBN: 0-07-137782-4

Ronanki, S. S. (1997). Response Analysis of Buried Circular Pipes under 3 Dimensional Seismic Loading, M.Tech thesis, Civil Engineering Department, Indian Institute of Technology, Kanpur, India

Susana Lopex Querol and Rafael Blazquez, (2006). Identification of failure mechanisms of road embankments due to liquefaction: optimal corrective measures at seismic sites. Canadian Geotechnic Journal, NRC Canada, 43: pp 889-0-902

Towhata, I. and Sim, W. W. (2010). Model Tests on Embedded Pipeline Crossing a Seismic Fault. ISBN: 978-984-8725-01-6, Proceedings of the 3rd International Earthquake Symposium, Bangladesh University of Engineering and Technology, Dhaka, 1-12

Unified Facilities Criteria. (2008). Structures to Resist the Effects of Accidental Explosions, UFC 3 340-02, Department of Defense, US Army Corps of Engineers, Naval Facilities Engineering Command, Air Force Civil Engineer Support Agency, United States of America

Zhenweng Yang. (1997). Finite element simulation of response of buried shelters to blast loadings, National University of Singapore, Republic of Singapore, International Journal of Finite Element in Analysis and Design, 24: pp 113-132

Permissions

The contributors of this book come from diverse backgrounds, making this book a truly international effort. This book will bring forth new frontiers with its revolutionizing research information and detailed analysis of the nascent developments around the world.

We would like to thank Prof. Abbas Moustafa, for lending his expertise to make the book truly unique. He has played a crucial role in the development of this book. Without his invaluable contribution this book wouldn't have been possible. He has made vital efforts to compile up to date information on the varied aspects of this subject to make this book a valuable addition to the collection of many professionals and students.

This book was conceptualized with the vision of imparting up-to-date information and advanced data in this field. To ensure the same, a matchless editorial board was set up. Every individual on the board went through rigorous rounds of assessment to prove their worth. After which they invested a large part of their time researching and compiling the most relevant data for our readers. Conferences and sessions were held from time to time between the editorial board and the contributing authors to present the data in the most comprehensible form. The editorial team has worked tirelessly to provide valuable and valid information to help people across the globe.

Every chapter published in this book has been scrutinized by our experts. Their significance has been extensively debated. The topics covered herein carry significant findings which will fuel the growth of the discipline. They may even be implemented as practical applications or may be referred to as a beginning point for another development. Chapters in this book were first published by InTech; hereby published with permission under the Creative Commons Attribution License or equivalent.

The editorial board has been involved in producing this book since its inception. They have spent rigorous hours researching and exploring the diverse topics which have resulted in the successful publishing of this book. They have passed on their knowledge of decades through this book. To expedite this challenging task, the publisher supported the team at every step. A small team of assistant editors was also appointed to further simplify the editing procedure and attain best results for the readers.

Our editorial team has been hand-picked from every corner of the world. Their multi-ethnicity adds dynamic inputs to the discussions which result in innovative outcomes. These outcomes are then further discussed with the researchers and contributors who give their valuable feedback and opinion regarding the same. The feedback is then collaborated with the researches and they are edited in a comprehensive manner to aid the understanding of the subject.

Apart from the editorial board, the designing team has also invested a significant amount of their time in understanding the subject and creating the most relevant covers. They scrutinized every image to scout for the most suitable representation of the subject and create an appropriate cover for the book.

The publishing team has been involved in this book since its early stages. They were actively engaged in every process, be it collecting the data, connecting with the contributors or procuring relevant information. The team has been an ardent support to the editorial, designing and production team. Their endless efforts to recruit the best for this project, has resulted in the accomplishment of this book. They are a veteran in the field of academics and their pool of knowledge is as vast as their experience in printing. Their expertise and guidance has proved useful at every step. Their uncompromising quality standards have made this book an exceptional effort. Their encouragement from time to time has been an inspiration for everyone.

The publisher and the editorial board hope that this book will prove to be a valuable piece of knowledge for researchers, students, practitioners and scholars across the globe.

List of Contributors

Feng Qing-Hai
State Key Laboratory of Disaster Reduction in Civil Engineering, Tongji University, Shanghai, China
CCCC Highway Consultants CO., Ltd. Beijing, China

Chih-Chen Chang
Department of Civil and Environmental Engineering, Hong Kong University of Science and Technology, Clear Water Bay, Kowloon, Hong Kong, China

Yuan Wan-Cheng
State Key Laboratory of Disaster Reduction in Civil Engineering, Tongji University, Shanghai, China

Akira Maekawa
Institute of Nuclear Safety System, Inc., Japan

Rehan Ahmad Khan
Aligarh Muslim University, India

Hasan Kaplan and Salih Yılmaz
Pamukkale University, Department of Civil Engineering, Turkey

Mohammad Reza Tabeshpour
Mechanical Engineering Department, Iran

Ali Akbar Golafshani
Civil Engineering Department, Sharif University of Technology, Tehran, Iran

Younes Komachi
Department of Civil Eng., Pardis Branch, Islamic Azad University, Pardis, Iran

Massimo Bavusi
CNR-IMAA, Italy

Vincenzo Lapenna and Antonio Loperte
CNR-IMAA, Italy

Romeo Bernini and Francesco Soldovieri
CNR-IREA, Italy

Antonio Di Cesare, Rocco Ditommaso and Felice Carlo Ponzo
Basilicata University/DiSGG, Italy

Genda Chen, Ying Huang and Hai Xiao
Missouri University of Science and Technology, Untied States of America

Mohamed F.M. Fahmy
Assiut University, Egypt

Zhishen Wu
Ibaraki University, Japan

A.J. Olarewaju, R.N.S.V. Kameswara and M.A. Mannan
Universiti Malaysia Sabah, Malaysia

Printed in the USA
CPSIA information can be obtained
at www.ICGtesting.com
JSHW011417221024
72173JS00004B/564